Body Composition Analysis of Animals

In recent years there have been substantial developments in the methodologies available for the non-destructive and non-invasive measurement of body composition in animals. By bringing together in a single volume a mix of traditional and well-established analytical methods with more modern techniques, *Body Composition Analysis: A Handbook of Non-Destructive Methods* provides a theoretical overview of different methodologies combined with practical advice on the use of these techniques. Methods covered include the use of destructive methods of analysis, body condition indices, isotope and gas dilution methods, total body electrical conductivity, bioimpedance analysis, ultrasound scanning and dual energy X-ray absorptiometry. Aimed at active research workers from undergraduate level upwards, this book will be of particular interest to those working in the fields of animal ecology, conservation biology, animal nutrition and physiology.

JOHN R. SPEAKMAN is Professor of Zoology at the University of Aberdeen, Scotland, and Head of the Division of Appetite and Energy Balance at the Rowett Institute, also in Aberdeen. He is a leading figure in the field of animal energetics. His research interests include the causes and consequences of variation in energy expenditure of a wide range of animals from bats and other small mammals to dolphins and turtles. In 1996 he received the Zoological Society of London Scientific Medal for his work on energetics and in 2000 was a Royal Society Leverhulme Senior Research Fellow.

Body Composition Analysis of Animals

A Handbook of Non-Destructive Methods

Edited by
JOHN R. SPEAKMAN

CAMBRIDGE UNIVERSITY PRESS
Cambridge, New York, Melbourne, Madrid, Cape Town, Singapore, São Paulo

Cambridge University Press
The Edinburgh Building, Cambridge CB2 8RU, UK

Published in the United States of America by Cambridge University Press, New York

www.cambridge.org
Information on this title: www.cambridge.org/9780521663380

© Cambridge University Press 2001

This publication is in copyright. Subject to statutory exception
and to the provisions of relevant collective licensing agreements,
no reproduction of any part may take place without the written
permission of Cambridge University Press.

First published 2001
This digitally printed version 2008

A catalogue record for this publication is available from the British Library

Library of Congress Cataloguing in Publication data
Body composition analysis of animals: a handbook of non-destructive methods / edited
by John R. Speakman.
p. cm.
Includes bibliographical references and index.
ISBN 0 521 66338 5
1. Body composition – Measurement – Handbooks, manuals, etc. 2. Non-destructive
testing – Handbooks, manuals, etc. I. Speakman, J. R. (John R.)
QP33.5 .B62 2001
572′.028′7 – dc21 2001018435

ISBN 978-0-521-66338-0 hardback
ISBN 978-0-521-08140-5 paperback

Contents

Contributors

MAURINE W. DIETZ
Centre for Ecological and Evolutionary Studies, Zoological Laboratory, University of Groningen, PO Box 14, 9750 AA Haren, The Netherlands

PETER R. EVANS
Department of Biological Sciences, University of Durham, Durham DH1 3LE, UK

JACK P. HAYES
University of Nevada, Reno, NV 89557, USA

BRIAN T. HENEN
Department of Biology, UCLA, Los Angeles, CA 90095-1606, USA

ELŻBIETA KRÓL
Aberdeen Centre for Energy Regulation and Obesity, University of Aberdeen, Aberdeen AB24 2TZ, Scotland, UK

THOMAS H. KUNZ
Department of Biology, Boston University, Boston, MA 02215, USA

P. IAN MITCHELL
Joint Nature Conservation Committee, Seabirds and Cetaceans, Dunnet House, 7 Thistle Place, Aberdeen AB10 1UZ, Scotland, UK

TIMOTHY R. NAGY
Division of Physiology and Metabolism, Department of Nutrition Sciences, Clinical Nutrition Research Center, University of Alabama at Birmingham, Birmingham, AL 35294-3360, USA

THEUNIS PIERSMA
Netherlands Institute for Sea Research (NIOZ), PO Box 59, 1790 AB Den Burg, Texel, The Netherlands

D. SCOTT REYNOLDS
Department of Biology, Boston University, Boston, MA 02215, USA

IAN SCOTT
Ecology Research Unit, De Montfort University, Lansdowne Road, Bedford MK40 2BZ, UK

COLIN SELMAN
Department of Zoology, Aberdeen Centre for Energy Regulation and Obesity, University of Aberdeen, Tillydrone Avenue, Aberdeen AB24 3TZ, Scotland, UK

J. SCOTT SHONKWILER
Department of Applied Economics and Statistics, University of Nevada, Reno, NV 89557, USA

JOHN R. SPEAKMAN
Department of Zoology, Aberdeen Centre for Energy Regulation and Obesity, University of Aberdeen, Aberdeen AB24 3TZ, Scotland, UK
and
Rowett Research Institute, Bucksburn, Aberdeen AB21 9BS, Scotland, UK

J. MATTHIAS STARCK
Institute of Systematic Zoology and Evolutionary Biology, Friedrich-Schiller-University, Erbertstaße 1, D-07743 Jena, Germany

WOUTER D. VAN MARKEN LICHTENBELT
Department of Human Biology, University of Maastricht, Maastricht, The Netherlands

GEORGE H. VISSER
Centrum voor Isotopen Ondezoek, Nijborgen, Groningen, The Netherlands

SALLY WARD
Bute Medical Buildings, School of Biology, University of St Andrews, St Andrews, Fife KY16 9TS, Scotland, UK

Preface

My first experience of writing a book was such that once it finally appeared I swore I would not write another. However, time heals, and after a few years had passed I entertained the notion of writing a second book, primarily because I was convinced that there was a need for a volume that summarized, in one place, useful information about non-invasive methods of body composition analysis. The problem, however, was that I wasn't really the best person to write the book because my own experience is limited to only a couple of the available methods. I decided therefore, that the best route would be to try and bring together a group of authors with the appropriate expertise and edit together their combined knowledge into a single text. I thought that, by editing a book, rather than writing it, I could still see the final volume realized, but that the process of producing it would be far less painful and stressful, and at the same time better by far than what I could achieve alone. And, indeed, this proved to be the case. But it was only like this because I had an excellent and co-operative group of colleagues on whom I could rely to deliver their chapters, and exceed my ambitious demands (that invariably clashed with examinations and teaching commitments) by only modest numbers of months.

Acknowledgements

My greatest thanks, therefore, go to all the authors of the individual chapters that make up this volume. Particular thanks are due to Matthias Starck, Maurine Dietz and Theunis Piersma who stood in at the last minute to replace another author who had to withdraw from the project on health grounds. (Hope you are feeling better Graham.) I think the final product is a credit to you all.

Thanks to those of you who put in time and effort (sometimes on multiple occasions) to peer review and improve the chapters that now comprise this book: Kim Hammond, Rob McAllen, Jack Hayes, Colin Selman, Sally Ward, Ian Mitchell, Ian Scott, Wouter van Marken Lichtenbelt, Brian Henen, Simon Heymsfield, Tom Kunz and Paul Thompson. I am also grateful to the anonymous reviewer selected by the publisher to provide an additional review of Chapter 5. Thanks are due to Tracey Sanderson of CUP who took the project on board and to the anonymous referees who favourably reviewed the original synopsis. I hope the final product lives up to the anticipations I had, and that you accepted as realistic.

While this book was being put together, Basil, our 15-year-old Cocker spaniel finally died – in good body condition to the very end.

J. R. Speakman
Aberdeen May 2000

JOHN R. SPEAKMAN

Introduction

Interest in the body condition of animals in the wild has long been an important aspect of animal ecology. Perhaps because there is an intuitive feeling that understanding something about the status of an animal – in terms of its body 'condition' – might provide a suitable window through which we can start to perceive components of an animal's fitness. The underlying premise behind this belief appears to be that an animal that is in 'good' condition is more likely to be a fitter member of the population than an animal that is in 'poor' condition. This notion may derive from the impression that the individual in 'good' condition has had the ability to not only satisfy all its requirements but also to take good care of itself. In addition, the animal may also have had time to deposit a healthy fat store that would see it through times of food scarcity. In contrast, the poor animal may be suffering from the ravages of a disease or simply be incapable of securing a living at the same time as managing to maintain itself. Intuitively therefore, by quantifying body condition, we may also be quantifying in a single measure these diverse factors that comprise our understanding of the term 'fitness'. Moreover, measuring the body conditions of the animals that live in a given area may not only tell us something about the animals but can perhaps also inform us about the area itself: its productivity, and the extent to which it can supply resources to sustain the population of animals that are living there.

More recently these notions have been added to by a body of work that has explicitly attempted to address the question of the adaptive nature of fat storage. In particular, these models have considered the level of fat that animals should store in their bodies (assuming they are not nutrient limited). The models predict that fat storage should be a responsive trait to fluctuations in the levels of predation along with the risks of starvation. When the risks of future shortfalls in energy supply increase, animals are predicted to elevate their body fat storage to enable them to ride out periods of difficulty. Conversely, if there is an increased risk of predation engendered by carrying around a large fat store, when predation risk is generally elevated animals might do better to carry around less fat.

Whatever the reasons, it is clear that there is a considerable and continuing demand among animal ecologists for methods that allow an assessment of the body condition (compositions) of their subject species. Probably one of the most popular methods for 'measuring' body composition of animals is to calculate some form of 'condition index'. The underlying premise of this method is the clear fact that individual animals are not all the same size. Hence, although measuring body mass gives a good indication of the total amount of tissue that an animal is carrying around, one might anticipate that bigger individuals would be heavier simply by virtue of their greater size. Overall body mass may be a poor measure of condition, but if one could somehow take size out of the equation, one would have an easily utilized method for 'measuring' animal body condition. The classical manner in which this has been done is to measure a linear aspect of the animals' body size (such as the length of its body) and then calculate a 'condition' index simply by dividing the mass by the length.

This was exactly the method I used in a paper in 1986 to express the body condition of immature bats and to try and relate spatial variations in this condition index (a) to differences in habitat quality around their roosts and (b) to the likelihood of the male individuals becoming sexually mature (Speakman and Racey, 1986). In that case we used body mass divided by the length of the bat's forearm to calculate a 'condition index'. Forearm length in bats is a repeatable and frequently used measure of 'size'. I think that most ecologists nowadays would be aware of the problems of scaling relationships and the difficulties that are attached to the derivation of such simple ratio based indices when the scaling is not isometric (although papers regularly still appear that ignore these problems) (see Packard and Boardman, 1987 for a full discussion). One might imagine that the difficulties attached to derivation of ratios are eliminated by utilizing more appropriate and sophisticated statistical approaches – such as regression analysis followed by the derivation of residuals. In their thought-provoking chapter in the first part of this book, Jack Hayes and Scott Shonkwiler (Chapter 1) show us the fallacy of relying on this reasoning and how even condition indices that appear to be derived using sophisticated statistical methods are, in fact, fraught with difficulties in derivation and interpretation.

What then is the animal ecologist expected to do? The obvious alternative is to resort to a more direct form of measurement, the most direct of which is chemical analysis. Chemical analyses of body composition have a long pedigree, particularly in the agricultural arena. Appropriate methods which minimize errors in the determinations have been worked out, and these have been summarized by Scott Reynolds and Tom Kunz in their chapter (Chapter 2). By adopting these methods, there is no doubt that the result is a precise and accurate breakdown of the body into its constituent aspects, providing in detail all the information an ecologist might need. Although this method beats the body condition index hands down when it comes to accuracy and precision, it has a clear drawback. The animal involved dies to give us the answer. Apart from the clear ethical concerns that such actions may pose, particularly where the subject species is rare or endangered, this action closes off the opportunity to explore the factors which influence variation in body composition within individuals over time.

One is therefore faced with a choice between two very different approaches: an approach that is completely non-invasive that may be dogged by problems of derivation and interpretation, or an ultimately invasive procedure that may be ethically dubious, incompatible with other aspects of the study, but precise and accurate. In the gap between these two alternatives are a group of methods that form the focus of the second part of this book. These are non-destructive methods that allow individual estimates of body composition, of varying accuracy and precision, but without the need to kill the subject to get the answer. However, before we get carried away on a wave of ethical self- congratulation, it is perhaps important to point out immediately that many animals do die in the process of validating these methods against the gold standard of chemical analysis. The methods may not therefore be quite as benign as they initially appear – a point that will be reiterated at several points throughout the text. Nevertheless, the methods do open up the possibilities of repeated measurements that are closed off by chemical analysis, and for most techniques they do provide a reduction in the degree to which animals are killed, compared with the numbers that might be destroyed if chemical analysis were the only alternative available.

The non-invasive methods included in this book fall into three natural groups. First, there are the methods that are based on some form of dilution principle. These include isotope-based methodologies that are the focus of my own contribution to the volume, which was co-authored by Henk Visser, Sally Ward and Ela Król (Chapter 3), and the lipid soluble gas methods explained by Brian Henen (Chapter 4). The second group of methods are those which exploit the electrical properties of body tissues to assess their composition – these include the chapter by Ian Scott, Colin Selman, Ian Mitchell and Peter Evans (Chapter 5), which concerns total body electrical conductivity (TOBEC) and the chapter by Wouter van Marken Lichtenbelt that covers bio-impedance analysis (BIA), a method used frequently in the clinical setting but increasingly being used with animals. Finally, the last two methods are based on the absorption and reflection properties of different tissues. These include a chapter by Matthias Starck, Maurine Dietz and Theunis Piersma on ultrasound imaging (Chapter 7) and finally a chapter by Tim Nagy on dual-energy X-ray absorptiometry (Chapter 8), recent machine developments of which have opened up tremendous opportunities for the study of small animals.

I am aware that, in selecting these methods, I have omitted several very powerful approaches to the non-destructive determination of body composition. These include, for example, computed tomography (CT) and magnetic resonance imaging (MRI) methods. I chose to omit these deliberately because, at present, these methods are effectively unavailable for the small animal ecologist, partly because the machines are generally prohibitively expensive, but also because even when they are available they are not portable, meaning animals have to be transported to the machines, rather than the reverse. In my mind, this sets them apart from the methods that are included in this volume because they are primarily laboratory methods rather than field methods.

Throughout the book we have attempted to standardize the use of a number of terms to avoid confusion. Body mass: is the total body mass including gut contents if present. Several different models have been used to subdivide the total body mass into separate components. These models have generally been called the two-, three- and four-compartment models reflecting the different numbers of compartments into which the body is subdivided. An unfortunate problem is that similar terms are used within each of these models to describe compartments that are not exactly equivalent. Hence the lean component of the body under the two-compartment model is not the same as the lean compartment under the

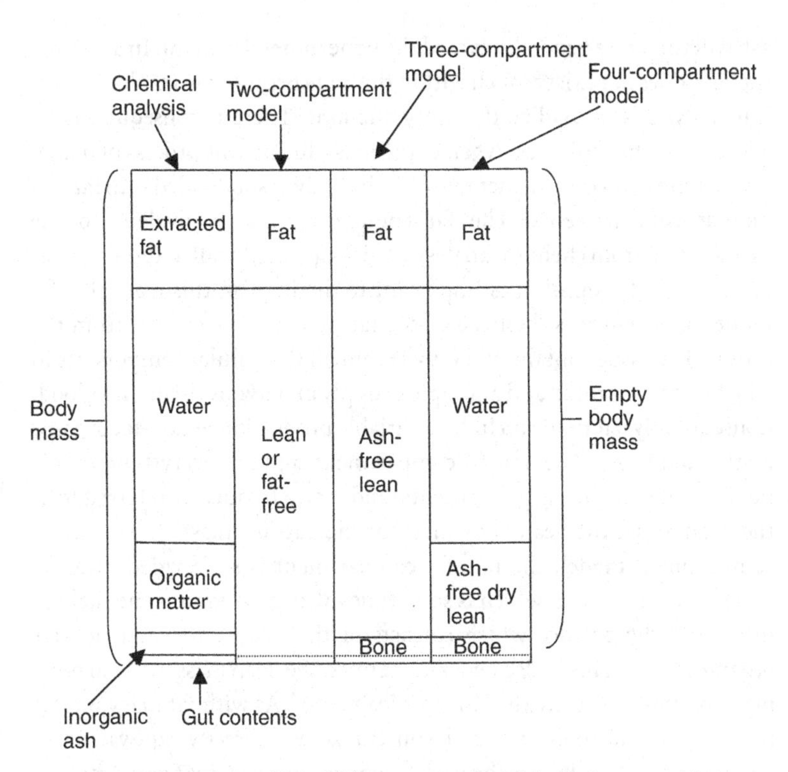

Fig. 1. Inter-relationships of different terms describing body composition analysis.

three- or four- compartment model. The subdivision of the body under different models is illustrated in Fig. 1, alongside the relationship of these divisions to the subdivision of the body when using chemical analysis. During chemical analysis, the gut contents are generally separated from the rest of the body. The mass of the remaining body is often called the 'empty body mass'. The mass of the gut contents can vary enormously, depending on the species under study. In ruminants, it can comprise a considerable portion of the total body mass. In animals that have been starving, they may be undetectable. The remaining body is separated into four constituents – extracted fat, water, organic matter and inorganic ash. Because the fat is extracted chemically, it is important to recognize that some of this fat will consist of structural lipids that are unavailable for utilization as an energy source by the animal in question. Inorganic ash derives principally from the skeleton. However, there is also a contribution from inorganic molecules spread throughout the body. Inorganic ash will therefore generally exceed the bone mineral content. In addition, the 'ash' does not exist as such within the body because the result of chemical analysis is to oxidize the inorganic ions. There is consequently an addition of atmospheric oxygen to the mass during this process of oxidation. In the two-compartment model, the body is subdivided into fat and non-fat compartments. The fat compartment is equivalent to the extracted fat from chemical analysis. This is generally called the 'fat mass' or the 'body fat mass'. It is important to distinguish this from the 'fat tissue mass' which is the mass of the fat tissues dissected out from the carcass. Fat tissue contains not only fat but all the cellular components in which the fat is stored and a complex supply of nervous tissue and blood. Consequently, fat tissue includes a variable proportion of non-fat organic matter and water. The non-fat compartment contains everything that is not fat (often including gut contents) and is also frequently referred to as the 'lean mass', the 'lean body mass' or the 'fat-free mass'. In the three-compartment model, the non-fat compartment is subdivided into the bone mineral content, which is sometimes also referred to as the skeletal mass, and the balance which consists of the body water and non-fat organic matter. This is also generally termed the 'lean mass' or 'lean body mass' or more correctly the 'ash-free lean mass'. As with fat mass and fat tissue mass, the bone mineral content is not directly equivalent to bone tissue mass, which is the mass of bones dissected out from a carcass. Bone tissue mass includes a considerable portion of organic tissue and water, above the mineral content. Finally, in the four-compartment model, the water content of the lean portion of the body is separated out – leaving the non-fat organic fraction, which is sometimes referred to as the 'dry lean mass' or 'ash-free dry lean mass'.

I selected authors for the chapters who have considerable practical experience with the methods that they have written about. This was important because each chapter contains not only a theoretical overview of the underlying principles by which each method works, but also invaluable practical advice on what to look out for when utilizing the methods in the field or laboratory. All the chapters were peer reviewed, normally by two independent referees who also had experience in the given field, in one case by only one referee, and for one chapter (TOBEC) by three referees, at the request of the publisher.

I sincerely hope you will find this book useful to guide you about decisions over what methods may suit your own study, and as a practical handbook during execution of any particular piece of work.

REFERENCES

Packard, G.C. & Boardman, T.J. (1987). The misuse of ratios to scale physiological data that vary allometrically with body size. In *New Directions in Ecological Physiology*, ed. M.E. Feder, A.F. Bennett, W.W. Burggren & R.B. Huey. Cambridge UK: Cambridge University Press.

Speakman, J.R. & Racey, P.A. (1986). The influence of body condition on sexual development of male brown long-eared bats (*Plecotus auritus*) in the wild. *Journal of Zoology (London)*, **210**, 515–25.

JACK P. HAYES AND J. SCOTT SHONKWILER

1

Morphometric indicators of body condition: worthwhile or wishful thinking?

Introduction

External measures of animal size, e.g. body mass, wing chord, foot length, and so on are often used by ecologists to develop body condition indices, e.g. body mass/length. Body condition indices are thought to reflect variation in diverse aspects of organismal quality, e.g. health, nutritional status, fat content and even Darwinian fitness (Krebs & Singleton, 1993; Brown, 1996; Viggers *et al.*, 1998). Body condition indices are generally easy to compute, so if they are highly correlated with variables, such as fatness or health, that are difficult to measure, they may be useful to ecologists for at least two reasons. First, they may be indicators of variables, e.g. fat content, that are difficult to measure accurately without harming an animal. Secondly, they may be more efficient or experimentally simpler to measure than variables that are hard to quantify, such as health or Darwinian fitness.

Ecologists seeking rapid non-destructive methods for estimating body condition have used two basic approaches for estimating body condition from external morphology. These two approaches are based on the construction of ratio variables, e.g. body mass divided by length, and the generation of residual variables, e.g. residuals from the regression of mass on length. Use of both kinds of condition indices requires considerable care to prevent errors in inferring the biology of interest, i.e. to prevent drawing erroneous conclusions about the biology being studied. One serious problem with the use of condition indices is that apparently subtle differences in the method used to compute the condition index may lead to vastly different conclusions about an animal's condition. A second problem is that researchers use condition indices to mean different things biologically. For example, condition indices are used as proxies for an animal's fat mass, health, survival, and so on. Thus the term condition index often is used imprecisely, which may lead to confusion in understanding what a condition index means. A third problem is that condition indices are sometimes used to capture biological concepts that are articulated imprecisely, e.g. well-being, health or quality. These problems lead to difficulties in evaluating the validity of condition indices in general. This chapter examines some of the commonly used condition indices based on external morphology. Our examination of the current usage of condition indices leads us to three conclusions. First, in many cases, the use of condition indices should be abandoned entirely. Secondly, when condition indices can be validated, and hence are still useful, greater care should be paid to the statistical analysis of condition indices formulated as either ratios or as residuals. Thirdly, we advocate the direct analysis of data used to generate condition indices as an alternative to the indirect, and often problematic, formulation of ratio or residual indices of condition.

What is body size?

Body size is an essential feature of condition indices, but the best measure of an animal's body size is a matter of debate (Rising & Somers, 1989; Freeman & Jackson, 1990; Piersma & Davidson, 1991; Wiklund, 1996). In reality, animals do not have a single size; they have many sizes (Bookstein, 1989). For example, a bird has a wet mass, a fat-free mass, a wing chord, a tarsus length, and a bill depth, all of which are measures of size. Which measure of size is the best probably depends on the biological question being asked. Yet, regardless of the size metric that is used, it is important to specify precisely what one means when referring to organismal size. A major feature of different measures of body size is that they may have different dimensions. For example, size can be assessed in dimensions of length, e.g. wing chord of a bird, snout–vent length of a lizard, standard length of a fish, total length of a mammal, and dimensions of mass. Throughout this chapter, unless specifically stated otherwise, we use the term mass to mean body mass. Mass is an indirect indicator of volume because an animal's mass equals its density multiplied by its volume. In geometrically similar objects, area increases in proportion to length squared, and volume in proportion to length cubed. If an object's density does not change with size, its mass increases in proportion to length cubed as well. The primary data for estimating body size are almost always measures of length or mass or some transformation of them.

Besides direct measures of size, such as mass, length, and so on, so-called structural size is an often used indirect measure of size, particularly for birds and mammals. Structural size is usually the first principal component (PC) from a principal components analysis of several measures of body length. Imagine that several measures of body length are available, e.g. bill length, bill depth, and tarsus length. A principal components analysis resolves the original *n* variables into *n* principal components so that (i) each PC score for an individual animal is a sum resulting from the linear combination of each of the original variables, e.g. $PC_1 = 0.498$ times bill length + 0.6 times bill depth + 0.7 times tarsus length, (ii) the principal components are all uncorrelated with one another, and (iii) the variance of the first PC is greater than the variance of the second PC and so on (Manly, 1986). If the first PC explains a large fraction of the cumulative variation in the lengths and if all the measures of length have positive loadings on the first PC, the first PC is often considered an overall metric of size and is called structural size. Clearly, an estimate of structural size depends on which body measurements are used to construct it, so again it is important to specify precisely how structural size is determined (Bookstein, 1989).

Condition indices and size

The condition indices, discussed in this chapter, are based on external measures of size. Two types of condition indices, ratios and residuals, are in common use. Ratio indices are one measure of size divided by a second measure of size. Ratio indices of condition include (i) mass divided by length, i.e. a linear measure of size; (ii) mass divided by length cubed; (iii) the cube root of mass divided by length; (iv) mass divided by length raised to an empirically determined power; (v) mass divided by mass predicted from length; (vi) log mass divided by log length; and (vii) log mass divided by log mass predicted from log length. Residual indices of condition are the difference between an observed measure of size and that predicted by a regression equation. Residual indices include (i) residuals from regressions of mass on length; (ii) residuals from regressions of mass on length cubed; (iii) residuals from regressions of mass on length raised to an empirically determined power; and (iv) residuals from regressions of log mass on log length. Hence, a multitude of alternative condition measures is available. While we will later contend that many, if not all, of these measures may be flawed, they are none the less widely used. Hence, for the present, we review the diverse ways in which condition indices are constructed and how these diverse condition indices are strongly dependent on the allometry or scaling of different body parts with organismal size.

Allometry and condition indices

Condition indices depend on body size, so the choice of a condition index critically depends on the associations between different measures of size. Of particular interest is the relationship between mass and length. In general, mass is strongly associated with length, so that much of the variation in mass is accounted for by length (Fig. 1.1). Despite this strong association, the relationship between mass and length is almost always non-linear. This non-linear relationship might well be expected from the dimensionality of mass and length. Dimensional arguments suggest that mass should vary not with length, but approximately with length cubed. While a perfect relationship between mass and length cubed might occur in geometrically simple objects such as a cube or a sphere, a perfect relationship is unlikely to occur in animals. In animals, as mass increases, shape and density may both change. Consequently, mass is unlikely to vary precisely with length cubed, and the precise scaling of mass with length must be determined empirically (Fulton, 1904). Two general questions are important to the determination of how mass scales with length. These questions are (i) what is the functional form of the relationship between mass and length on average, and (ii) how do individuals deviate from the overall relationship?

Form of the mass to length relationship

The relationship between mass and length is usually modelled as a special case of the equation:

$$\text{mass} = \delta + \alpha\,\text{length}^{\beta}.$$

This general equation suggests that mass is a power function of length plus an intercept term. The parameters of this general equation can be estimated by non-linear regression, but to the best of our knowledge such models are rarely fitted by ecologists (cf. Albrecht *et al.*, 1993; Hayes *et al.*, 1995). Rather than fitting the general model, biologists sometimes, but not always, implicitly assume that the intercept term is zero and hence that $\text{mass} = \alpha\,\text{length}^{\beta}$. The assumption of a zero intercept seems intuitively

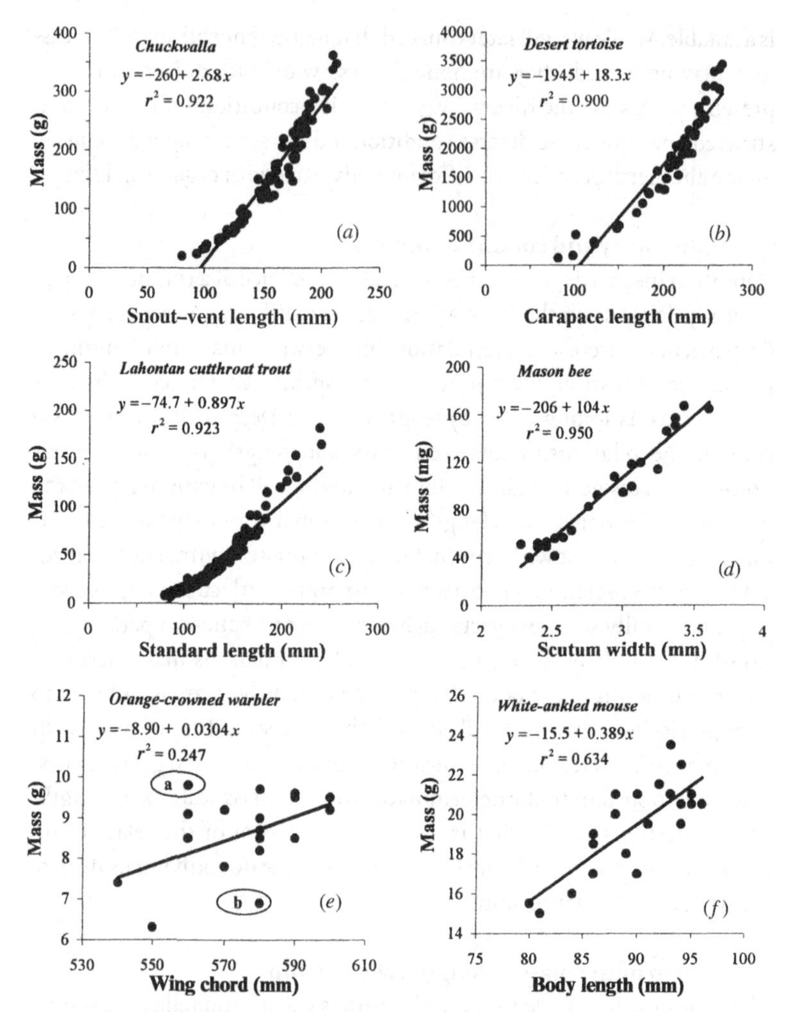

Fig. 1.1. Bivariate plots of mass on length for six diverse taxa. Linear regressions of mass on length are shown. The original data were obtained from the authors listed in the acknowledgements. In a few cases, outliers that are not shown were deleted from the original data sets.

reasonable because animals with zero length ought to have zero mass. One way to determine β is to perform a non-linear regression, but biologists rarely determine scaling relationships in this way. Instead, the standard procedure is to determine β from log transformed data. Log transforming the power function without an intercept, i.e. log transforming the equation: mass $= \alpha$ length$^{\beta}$ produces a relationship that is linear in the logarithms of the data, i.e. log mass = log $\alpha + \beta$ log length. It is important to recognize that the parameters $\alpha + \beta$ determined by the non-linear regression generally do not equal the parameters $\alpha + \beta$ determined by the log-log regression (Zar, 1968; Smith, 1993). Consequently, the two models (log mass = log $\alpha + \beta$ log length and mass = α length$^{\beta}$) are related but not identical. Choosing between these models for the scaling of mass to length requires that, in addition to examining the functional form of the relationship, the residual (error) variation must be considered.

Individual variation, i.e. error variance and the scaling of mass to length

As a general rule, mass will not vary exactly with any simple function of length, rather individual animals will deviate from the overall pattern. These individual deviations are hoped to be indicative of animal condition. It is possible to arbitrarily assume the nature of the relationship with mass and to fit the best descriptive relationship without specifying the distribution of errors (residuals). For example, the least-squares regression of mass on length can be fitted without specifying an error distribution. However, in the absence of assumptions about the error distribution, standard approaches to statistical inference cannot be used. Consequently, because biologists generally want to make inferences from their studies, reasonable assumptions about the error distribution generally must be made.

While not the only valid approach, the statistical methods used by biologists typically are based on tests which assume that errors are normally distributed with equal variance. If the residuals from a regression analysis are not consistent with this assumption, either remedial measures are in order or alternative model specifications are needed. The significant point is that model fitting requires analysis of both functional form and the error distribution.

To illustrate the importance of accounting for error distributions, consider choosing between fitting a power model, i.e. mass = $\delta + \alpha_a$ length$^{\beta_a}$, and fitting a linear model to the logarithms of the data, i.e. log mass = log $\alpha_m + \beta_m$ log length. Hereafter, the subscript **a** refers to estimates obtained on the arithmetic scale, and the subscript **m** refers to estimates on the logarithmic scale. In general, the slope of the log–log regression (β_m) does not equal the exponent (β_a) on the arithmetic scale. The αs also will differ. In the power model, residual variation in mass is minimized. In the log–log model, residual variation in log mass is minimized. Which model is more appropriate to fit depends on the distribution of the errors. Suppose that ε is normally distributed. If mass $= \alpha$ length$^{\beta} + \varepsilon$, the non-linear least-squares model is appropriate because the errors are additive and normally distributed. If mass $= \alpha$ length$^{\beta}$ e$^{\varepsilon}$, where e is the base of the logarithms used to transform the power function to a linear form, the log–log regression is appropriate because the errors on the arithmetic, i.e. not logarithmic scale are multiplicative and lognormally distributed. The logarithm of a variable with a lognormal distribution is normally distributed. A major point is that choosing which model is appropriate depends on the error distribution and whether the errors are additive (mass $= \alpha$ length$^{\beta} + \varepsilon$) or multiplicative (mass $= \alpha$ length$^{\beta}$ e$^{\varepsilon}$).

In summary, allometric models that scale mass to length are central to indices of condition. Even indices of condition that are not explicitly based on allometric models are affected implicitly by the scaling of mass and length. Hence, allometric reasoning is essential to a discussion of condition indices.

Simple ratio indices of condition

How are measures of size used to formulate condition indices? The simplest body condition index is a ratio of sizes, for example, mass divided by length, where the anatomical length chosen depends on the animal being studied. In birds, length has been measured as wingspan, tarsus length, and bill depth; in fish, as fork length and standard length; in mammals, as jawbone length, body length excluding tail, and body length including tail; in squamates, as snout–vent length; in turtles, as carapace length, and so on. Instead of a single measured length, structural size may also be used as the denominator of a condition index. The assumption behind these ratio formulations of condition indices is that, for any given length (or structural size) animals that are heavier, i.e. have greater mass, must be fatter or have more bone, muscle and other tissue. Animals that are heavier for a given length are presumably in better condition. In wild animals, where the acquisition of resources is central to survival and reproduction, i.e. Darwinian fitness, animals with greater fat or other body tissue reserves may be more likely to survive or reproduce than animals with lower reserves.

The motivation behind the use of condition indices can be seen by examining the relationship between mass and length in diverse taxa (Fig. 1.1). For any given length of animal, mass will vary. The amount of variation about the length varies across taxa. For example, the variation in mass for any given length is substantial in the orange-crowned warbler (*Vermivora celata*) and in the white-ankled mouse (*Peromyscus pectoralis*), but is much less for chuckwalla (*Sauromalus ater* – a lizard), the desert tortoise (*Gopherus agassizii*), the cutthroat trout (*Oncorhynchus clarki*), and the mason bee (*Osmia lignaria*). A simple ratio index of condition (mass/length) is an attempt to summarize the variation in mass relative to length. For example, the individual labeled *a* in Fig. 1e has a mass of 9.8 g and a length of 560 mm. Its condition index is 0.0175. Similarly, the individual labeled *b* in Fig. 1e has a mass of 6.9 g and a length of 580 mm. Its condition index is 0.0119. The hope is that this condition index will provide a reliable estimate of fatness or nutritional status, such that bird *a* is fatter or in better condition than is bird *b*. Unfortunately, this approach has serious potential flaws.

Many ratio condition indices are not size independent

If the goal of a condition index is to compare fatness or nutritional status or some other metric of organismal quality across animals, a simple ratio index may not accomplish this goal acceptably. A major problem is that the condition index almost certainly is correlated with size. Correlations with size are a problem because a major objective of condition indices often is to compare the condition of animals with different sizes (Cattet, 1990; Jakob *et al.*, 1996). If condition indices are correlated with size, animals that differ in size would be expected to have different condition indices just because they differ in size. For example, compare a small and a large chuckwalla (Fig. 1.1(*a*)), where the large (200 mm snout–vent length) lizard is twice as long as the small (100 mm snout–vent length) one. Each animal has the predicted mass for lizards of that length, but compared to each other, one lizard will have a larger condition index, because the condition index is correlated with size. The dependence of condition index on size is based on the underlying relationship between mass and length, which often is well described by a power function (Fig. 1.2). Suppose that the relationship is such that mass $= \delta + \alpha$ length$^{\beta}$. If we ignore the effects of individual deviations from the overall relationship and if the power function fits the data, the relationship between the condition index (mass/length) and length must be: condition index = mass/length $= (\delta + \alpha\ \text{length}^{\beta})/\text{length}^{1} = \delta/\text{length} + \alpha\ \text{length}^{(\beta-1)}$. In other words, condition index will depend on length except when β equals one

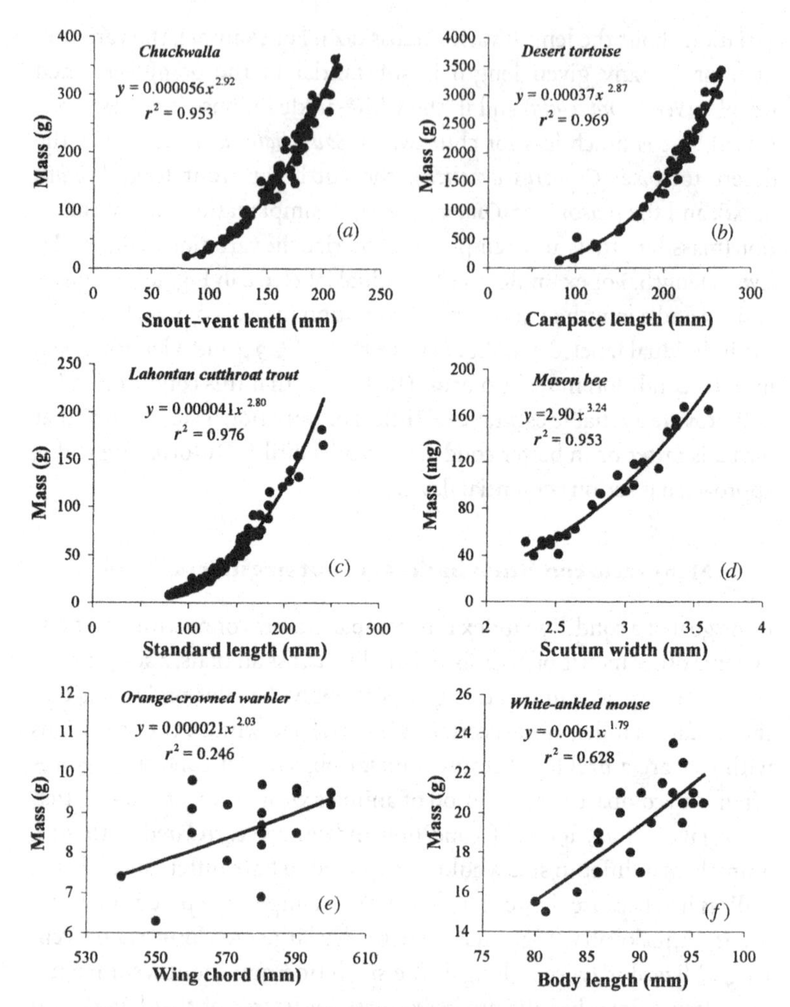

Fig. 1.2. Bivariate plots for the same data as depicted in Fig. 1.1. Non-linear regressions with no intercept are shown. The corresponding equations obtained from a log–log model are: (*a*) mass = $0.000021\,x^{3.11}$, (*b*) mass = $0.00086\,x^{2.71}$, (*c*) $y = 0.000015\,x^{3.00}$, (*d*) mass = $2.56\,x^{3.35}$, (*e*) mass = $0.000086\,x^{2.17}$, and (*f*) $y = 0.0074\,x^{1.51}$.

and $\delta=0$, which is extremely unlikely. Returning to the lizard example, assume that the power function with $\alpha=0.000056$ and $\beta=2.92$ fits perfectly and that $\delta=0$, so that the condition index $=0.000056$ length$^{1.92}$. The condition index is 0.39 for the smaller lizard and 1.47 for the larger lizard. Both animals are completely typical for their length, but the large lizard has a condition index almost four times that of the small lizard. This condition index seems flawed because it inherently incorporates an association between length and condition. Hence, ratio condition indices based on mass/length seem unlikely to be effective tools for comparing the condition of animals that vary in size.

Despite the preceding example, it is possible to argue for the size independent use of simple ratio condition indices (mass/length) in one special case. The special case is when the relationship between mass and length is a power function with the intercept (δ) equal to 0 and the exponent (β) equal to 1. When this special case applies, then (i) the relationship between mass and length is a line that passes through the origin and has slope of α and (ii) the condition index is α length$^{(\beta-1)}=\alpha$ length$^{0}=\alpha$, which obviously does not vary with size, i.e. length. The condition index equals α for every animal that has the predicted mass for its length.

Correcting ratio indices for differences in dimension

Fisheries biologists noted long ago that mass is proportional to density times length cubed (Fulton, 1904; Le Cren, 1951). Hence, condition indices of the form (mass/length) should be roughly a function of length squared, if density does not change with size (Fulton, 1904). One way to attempt to remove the length dependence of this relationship is to calculate a condition index as mass/length^3. This condition index typically has a small value because the denominator is cubed, hence it may be rescaled by multiplying by some power of 10. This rescaled number is called Fulton's condition index (Ricker, 1975*b*). Fulton's condition index is used primarily in the fisheries literature, but analogues to it are found in studies on diverse taxa (Dunn & Chapman, 1983; Jorgensen, 1983; Tuomi & Jormalainen, 1988; van Marken Lichtenbelt *et al.*, 1993; Johnsen *et al.*, 1996; Droney, 1998). Ignoring individual deviations for the moment, suppose that the actual relationship between mass and length is mass $=\delta+\alpha$ length$^{\beta}$. Fulton's condition index is then $k\delta$/length$^3+k\alpha$ length$^{\beta-3}$, where k is the rescaling factor. Hence, Fulton's condition index will depend on size (length^3) unless $\delta=0$ and $\beta=3$. Such a situation is likely to be rare. Thus, as for mass/length, mass/length^3 is likely to be correlated with size in most data sets (cf. Cone, 1989). When the goal is to generate a condition index that is independent of size, use of Fulton's condition index may be reasonable if mass scales with length cubed or nearly so (and δ is close to 0).

Building ratio indices with an allometric scaling exponent

The data sets in Fig. 1.2 are consistent with reports which show that mass typically scales to a power of length other than one. While the scaling of mass with length is often close to three, appreciable deviations from this value (three) are common. If the scaling exponent is not reliably one or three, obtaining a size-independent (where size is length or length raised to a power) condition ratio requires that some scaling exponent other than one or three be used to form the condition index.

Normal, additive errors and non-linear regression

The proper method for estimating the scaling exponent of mass on length depends on the nature of the error variance. If the errors are additive and normally distributed, a non-linear regression model is appropriate. Suppose that you fit the non-linear regression model, mass $= \delta + \alpha$ length$^{\beta} + \varepsilon$. The condition index is mass ($\delta + \alpha$ length$^{\beta} + \varepsilon$) divided by length$^{\beta}$, which equals $\alpha + \delta$/length$^{\beta} + \varepsilon$/length$^{\beta}$. When $\delta \neq 0$ or when mass is divided by length raised to an exponent other than β, the condition index will be not size independent. This result is the reason why mass/length and Fulton's condition index are typically correlated with size. In contrast, when $\delta = 0$ and the condition index is calculated by dividing by length$^{\beta}$, the expected correlation between the condition index and size (length$^{\beta}$) will be zero, i.e. at least within the limits of a second-order Taylor series expansion (Mood *et al.*, 1974). In practice, owing to sampling variation, the actual correlation will not exactly equal zero, but it should be close to zero at least in large samples.

Why is the expected correlation zero, when $\delta = 0$? When $\delta = 0$, the condition index is $\alpha + \varepsilon$/length$^{\beta}$. The first term α is a constant, so it is uncorrelated with length$^{\beta}$. By design, regressions are fitted so that ε (the residuals) and the predictors, i.e. length$^{\beta}$ will be uncorrelated and that ε will have a mean of zero. When ε and length$^{\beta}$ are uncorrelated and ε has a mean of zero, a second-order Taylor series expansion shows that the expected correlation of ε/length$^{\beta}$ and length$^{\beta}$ is zero (Mood *et al.*, 1974; cf. Pearson, 1897; Atchley *et al.*, 1976). Hence, the expected correlation between the ratio index and length$^{\beta}$ will be zero. Thus, whenever we fit the scaling relationship by regression and divide by length raised to the empirically determined scaling exponent, the condition index is size (length$^{\beta}$) independent.

When $\delta \neq 0$, a size-independent condition index can be formulated by dividing mass by predicted mass, i.e. calculate condition as $(\delta + \alpha\ \text{length}^{\beta} + \varepsilon)/(\delta + \alpha\ \text{length}^{\beta})$, which equals $1 + \varepsilon\ /(\delta + \alpha\ \text{length}^{\beta})$. The second term has an expected correlation of zero with length$^{\beta}$, so the index has an expected correlation of zero with length$^{\beta}$. In practice, owing to sampling variation, the actual correlation will not exactly equal zero, but it should be close to zero at least in large samples.

Lognormal, multiplicative errors and log–log regression

When the errors (e^{ε}) are multiplicative, i.e. mass $= \alpha\ \text{length}^{\beta}\ e^{\varepsilon}$ and log-normally distributed, log–log regression should be used instead of non-linear regression. If a $\log_e$–$\log_e$ model is fitted, the condition index may be computed on either the arithmetic scale, i.e. mass/length$^{\beta}$, or the logarithmic scale, i.e. $\log_e$ mass/ $\log_e$ length. The arithmetic condition index ($\alpha\ \text{length}^{\beta}\ e^{\varepsilon}/\text{length}^{\beta} = \alpha\ e^{\varepsilon}$) can be calculated using the exponent determined from the slope (β) of the log–log regression. This condition index has an expected correlation of zero with size (length$^{\beta}$). The reason the expected correlation is zero is as follows. Because the relationship was fitted by log–log regression, ε and $\beta \log_e$ length are uncorrelated. Now raise e to the ε and $\beta \log_e$ length to obtain e^{ε} and length$^{\beta}$. These transformed variables (e^{ε} and length$^{\beta}$) can be shown to have an expected correlation of zero (Hogg & Craig, 1995), so the index and length$^{\beta}$ have an expected correlation of zero. In practice, owing to sampling variation, the actual correlation will not exactly equal zero, but it should be close to zero at least in large samples.

In contrast, the logarithmic scale condition index is generally not size independent. The logarithmic scale condition index ($\log_e$ mass/$\log_e$ length) equals ($\log_e \alpha + \beta \log_e$ length $+ \varepsilon$)/($\log_e$ length) which simplifies to $\beta + \log_e \alpha/\log_e$ length $+ \varepsilon/\log_e$ length. Hence, except when the intercept ($\log_e \alpha$) equals zero, the condition index calculated with log transformed data will depend on size, i.e. $\log_e$ length, because $\log_e \alpha/\log_e$ length varies with size. A size-independent condition index can be calculated with log transformed data if, instead of dividing log mass by log length, log mass is divided by predicted log mass. That is, calculate condition by dividing log mass by predicted log mass, i.e. the condition index is $(\log_e \alpha + \beta \log_e \text{length} + \varepsilon)/(\log_e \alpha + \beta \log_e \text{length})$, which equals $1 + \varepsilon/(\log_e \alpha + \beta \log_e \text{length})$. Following the reasoning above, this condition index has an expected correlation of zero with size ($\log_e$ length).

Ratio indices – summary

Some ratios indices, e.g. mass/predicted mass have an expected correlation of zero with size, but most ratio indices, e.g. mass/length and mass/length^3 and log mass/log length are correlated with size. These size-dependent ratio indices can be misleading because the effects of the condition index may be confounded with the effects of size. For example, if both size and condition index are correlated with number of offspring, a correlation between the condition index and number of offspring may simply be due to their mutual correlation with size. Besides this problem, numerous potential statistical and inferential pitfalls plague the use of ratio variables (Bollen & Ward, 1979; Kronmal, 1993, 1995; Allison *et al.*, 1995; Beaupre & Dunham, 1995; Jasienski & Bazzaz, 1999; Packard & Boardman, 1999 and references therein, but see Jungers *et al.*, 1995; Neville & Holder, 1995 for counterarguments). Consequently, we recommend caution when using ratio indices.

Residual indices of condition

Unlike the situation with ratios, it is easy to obtain size-independent condition indices with residuals from regressions. For example, suppose that mass is regressed on the total body length of mice (Fig. 1.1(*f*)). A mouse whose mass falls above the regression line has a higher mass for its length than a mouse whose mass falls below the regression line. If the mouse's condition is assessed by its regression residual, i.e. its actual mass minus that predicted for a mouse of that length by the regression, the mouse with the largest positive residual is in the best condition and the mouse with the largest negative residual is in the worst condition.

A major benefit of residual indices of condition is that they are uncorrelated with size. This lack of correlation with size is an advantage because it helps prevent confounding the effect of size on other variables with the effect of condition on other variables (cf. Hayes & Shonkwiler 1996). Specifically, the residual index is uncorrelated with whatever measure of size is the predictor on which mass or log mass is regressed.

This property is the result of the standard methods used to calculate ordinary least-squares (OLS) regressions with an intercept. By design, an OLS regression is estimated so that the residuals, e.g. condition and the predictor(s), e.g. length are uncorrelated. Hence, the condition index and the size predictor of it will be uncorrelated if the residual index is formed by ordinary least-squares regression. Incidentally, the residuals (condition) and the dependent variable (mass, in our example) must be correlated except in the trivial case when mass (the dependent variable) is perfectly correlated with the predictor (Draper & Smith, 1981; Framstad *et al.*, 1985).

In practice, residual-based estimates of condition are often generated by regressing mass on length. Nonetheless, analogous to the situation for ratio indices, residual indices may be based on other models. For example, we can estimate residuals in at least five ways: (i) residuals from a regression of mass on length; (ii) residuals from a regression of mass on length cubed; (iii) residuals from a regression of log mass on log length; (iv) residuals from an empirically fitted power function without an intercept; and (v) residuals from an empirically fitted power function with an intercept. All these models produce residual indices that are uncorrelated with size. This size independence is a possible advantage over many ratio indices.

Despite the indices' size independence, investigators are still faced with choosing from a number of alternative models for generating residuals. This choice is not trivial. Inspection of the residual condition indices for some of the alternative regressions makes it clear that the condition indices are affected in a major way by the choice of regression used (Fig. 1.3, Table 1.1). Statistical criteria may be used to help choose between the alternatives, but even if a particular regression model fits well, that does not guarantee that the index will be a good indicator of the relevant biology. Regardless of statistical criteria, an index may still be a poor indicator of fat mass, survival, etc. Later, we discuss this point in more detail.

Residuals from regressions other than OLS

While many researchers have constructed condition indices from OLS regressions on length or some transformation of length, yet another approach to generating residuals is to use alternative regression models (Jakob *et al.*, 1996). The most commonly advocated alternative to OLS regression is the reduced major axis (RMA) regression. The argument for using RMA regression is that OLS depends on an assumption that is often untrue, i.e. that the predictor values are measured without error. Whenever a predictor contains error, OLS is an inappropriate statistical

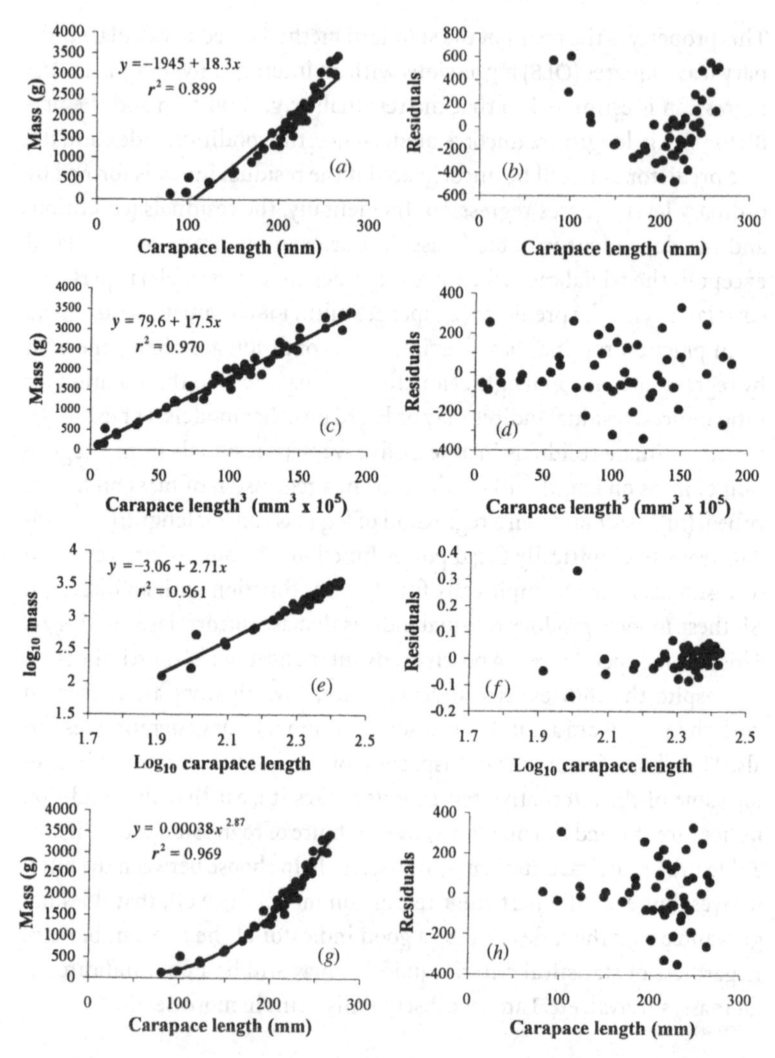

Fig. 1.3. Four regression models and the residual condition indices they imply for the desert tortoise data shown in Figs. 1.1 and 1.2. The different regression models result in very different condition indices (Table 1.1).

Table 1.1. *The correlation between various residual (ε) indices of condition for the desert tortoise data*

	Linear	CL³	Log–log	Non-linear – no intercept	Non-linear – with intercept
Linear	1	0.570	0.494	0.623	0.541
CL³		1	0.699	0.994	0.999
Log–log			1	0.701	0.695
Non-linear – no intercept				1	0.990
Non-linear – with intercept					1

Notes:
The column and row labels refer to the following regressions: linear (mass = $\delta + \beta$ carapace length + ε), CL³ (mass = $\delta + \beta$ carapace length$^3 + \varepsilon$), log–log (log mass = $\delta + \beta$ log carapace length + ε), non-linear–no intercept (mass = α carapace length$^\beta + \varepsilon$), and non-linear–with intercept (mass = $\delta + \alpha$ carapace length$^\beta + \varepsilon$).

model and an alternative model may more accurately represent the underlying relationship in the data uncontaminated by error. An RMA regression is one of an infinite number of alternatives to OLS regression. These alternatives to OLS are all distinguished by the particular assumptions that they make about the magnitude of error variances (see Rayner, 1985; Harvey & Pagel, 1991). The OLS slope is fitted based on the assumption that the error variance of the predictor variable is zero. The RMA slope is fitted based on the assumption that the ratio of the error variance to the true variance is the same for both the variables in the regression. Other assumptions about the error variance of the predictor lead to slope estimates different from those estimated by OLS or RMA.

The debate over which regression is appropriate is a complex one (Ricker, 1973, 1975*a*; Jolicoeur, 1975; McArdle, 1988; LaBarbera, 1989). The key issue is that biological variables often contain measurement error. When a predictor variable contains measurement errors, the OLS slope is biased. In a simple regression, if the predictor contains measurement error, the OLS slope will be attenuated, i.e. slope closer to zero than the true slope, but the significance test for whether the slope is different from zero will still be correct (Fuller, 1987). When there is more than one predictor, the effects on slope estimates and significance tests are complex (Fuller, 1987), even if only one predictor is measured with error. RMA is the most commonly used approach other than OLS, so our comments on alternatives to OLS are restricted to RMA.

The slope fitted in an RMA regression with a single predictor is equal to the ratio of the standard deviations of the variables in the regression. For example, the slope for generating an RMA of mass on length equals the standard deviation of mass over the standard deviation of length. The RMA slope also equals the default OLS slope divided by the correlation between the dependent and independent variables. Hence, unless the correlation between mass and length is perfect, the RMA has a steeper slope than the OLS slope, i.e. the OLS slope is attenuated. Let us assume that the RMA is the appropriate statistical model for data to be used in generating a residual index of condition. The residual condition indices from an RMA will differ in several ways from the residuals (condition indices) from an OLS regression (Fig. 1.4). First, the slope is steeper, although both regressions pass through the bivariate mean of the data. Consequently, the predicted mass for all animals whose length is less than the mean will be lower than for the OLS regression, and the predicted mass for all animals whose length is greater than the mean will be greater than for the OLS regression. The residual is the observed value minus the predicted value, so small animals (with decreased predicted values) will tend to have higher condition indices in a condition index derived from an RMA (Fig. 1.4). Likewise, large animals (with increased predicted values) will tend to have lower condition indices in a condition index derived from an RMA. Another difference from the case with OLS regression is that the RMA-derived residuals will be correlated with the measured lengths (or transform of length), so that the index will not be size-independent (Harvey & Pagel, 1991). In the orange-crowned warbler example shown in Fig. 1.4, the residuals from the RMA are negatively correlated ($r = -0.501$) with wing chord. However, if the RMA accurately accounts for the measurement error in the lengths, the true lengths, i.e. the observed lengths minus the measurement errors will be uncorrelated with the residuals, i.e. the condition index, even though the observed lengths are correlated with the residuals (Fuller, 1987; Bollen, 1989).

In conclusion, whether it is appropriate to use OLS regression, RMA analysis, or some other statistical model to generate residual indices of condition, depends on the error variances (and if present, error covariances) of the data under study. Biologists rarely attempt to account for measurement error variance empirically. Instead, the choice of the appropriate regression model is usually made arbitrarily. In practice, it is often difficult or impossible to estimate what those error variances and covariances might be, so despite their generally unsubstantiated validity,

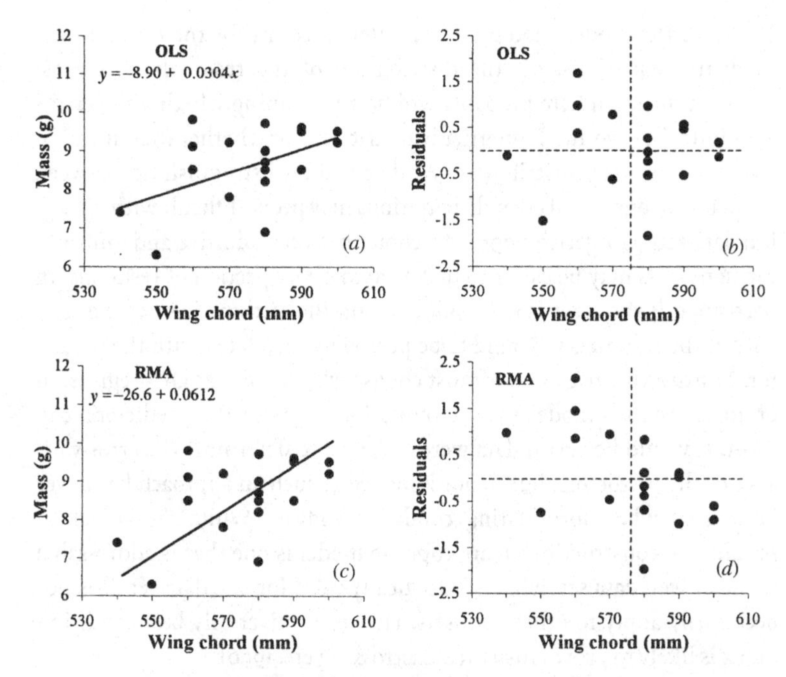

Fig. 1.4. A comparison of OLS and RMA slopes and residual condition indices computed from them for the orange-crowned warbler data shown in Figs. 1.1 and 1.2. The RMA slope is steeper than the OLS slope, but both the OLS and RMA lines pass through the bivariate mean. Hence, residual condition indices computed from the RMA are higher for small birds, i.e. those smaller than the mean because the RMA predicts lower mass for these birds than OLS does. Residual condition indices computed from the RMA are lower for large birds, i.e. those bigger than the mean because the RMA predicts higher mass for these birds than OLS does. The dashed lines in (*b*) and (*d*) intersect at the bivariate mean for the residuals and wing chord.

condition indices are most likely to continue to be based either on OLS or RMA models.

Statistical criteria for fitting mass to length relationships

When using regression as a precursor to building a condition index, the appropriate statistical model to use is not something that can be specified *a priori*. The best-fitting model needs to determined empirically. Model specification is a non-trivial subject, but for the current purposes there are three issues. First, an appropriate functional form needs to be specified, e.g. should mass be regressed on length or mass regressed on length^3, etc.

Secondly, the model should appropriately account for the errors in the predictive equations, i.e. the distribution of the residuals. The most common question here probably will be determining whether the errors are multiplicative and lognormally distributed or whether they are additive and normally distributed. While reliably distinguishing between normal and lognormal error distributions may prove difficult with biological data sets (Gingerich, 1995), the choice between additive and multiplicative models may be easier to make based on inspection of residuals. In particular, if the error is multiplicative, on the arithmetic scale the variance of the residuals will tend to be positively correlated with the predictor. Thirdly, the investigator must choose whether to use OLS regression or an alternative model that accounts for errors in the predictors, e.g. RMA. It would be best if data were collected to determine the error variance of the predictors, but to our knowledge such an approach has never been used when formulating condition indices. While the statistical problem of constructing an appropriate model is one that is addressable for any given data set, what is statistically valid for one data set does not necessarily apply to other data sets. Hence, a universally best condition index is likely to prove elusive (cf. Castro & Myers, 1990).

What condition indices are most commonly used?

The number of studies that have used condition indices is large (>1000), and we have not attempted to review them comprehensively. Rather, we performed a limited search of the available studies to obtain an indication of what kinds of condition indices are most commonly used. To do so, we searched the online databases of *Bioabstracts* and the *Science Citation Index*. Initial efforts produced numerous papers on fish, fewer papers on birds, even fewer on mammals, and fewest of all for other taxa. Hence, we searched by major taxon to sample the use of condition indices across a broad spectrum of taxa. We ignored studies of condition in fish because the most common condition index (~ 50% of all studies) used by fisheries biologists already is known to be the ratio of mass divided by length^3 (Bolger & Connolly, 1989). We examined the indices used in 22 studies of mammals, 21 studies of birds, 7 studies of reptiles and amphibians, and 8 studies of invertebrates. Ratio indices were used in 31 studies, residual indices were used in the other 25, and hybrids of residuals divided by predicted mass were used in two studies. Mammalogists used ratio indices more commonly ($n=16$) than they used residual indices ($n=6$). Ornithologists used residual indices more commonly ($n = 13$) than they used ratio indices ($n = 6$). The two hybrid indices were constructed by ornithologists (Hood *et al.*, 1998; Royle & Hamer, 1998). Across all the studies examined, a wide variety ($n = 16$) of different indices was used.

Should ecologists use condition indices?

Despite the complexities associated with appropriately modelling mass to length relationships and the wide diversity of indices that are used in practice, the greatest concern with condition indices is not statistical; it is conceptual. Sometimes, condition indices are used as indicators of fat content or to summarize mass to length relationships. But, often the intended biological meaning of a condition index is nebulous. Typically, the aspects of health or nutritional status or robustness or organismal quality that condition indices ostensibly summarize are not stated precisely, and their validity is not verified. Even when a condition index is intended to describe something as simple as fat mass, fat mass may turn out to be a poor indicator of nutritional status (Batzli & Esseks, 1992). In the absence of additional data, the use of condition indices without validation is little more than wishful thinking. It is crucial that the validity of a condition index be established (cf. Krebs & Singleton, 1993; van der Meer & Piersma, 1994). Otherwise, the condition index may not indicate exactly what the biologist using it thinks it indicates.

We compare one of many possible residual indices with one of many possible ratio indices to illustrate the problem of establishing the validity of a condition index. The measure of validity that we consider is the criterion validity, i.e. the correlation between the indicator and what it is intended to measure (Bollen, 1989). Suppose that mass is a linear function of length, i.e. mass $= \delta + \beta$ length $+ \varepsilon$, and that a short fish, a medium fish, and a long fish each has a residual condition index of 10 g. All have the same above-average condition, because they all have the same residual. However, it may be that a short fish with a residual of 10 g has a huge energy reserve, and hence a large advantage in surviving a food shortage. In contrast, the residual 10 g is a smaller fraction of body mass in the long fish, and consequently it may confer relatively little benefit during a food shortage. In other words, being 10 g heavier may be a greater advantage to a small fish than it is to a large fish. If this is true, the residual measure of condition may not be indexing the relevant biological variable, i.e. survival during food shortage. Alternatively, suppose that a ratio index of condition, i.e. mass/predicted mass is used and that the predicted masses of the three fish are 20, 100, and 500 g. Then the ratio indices of condition would be 1.50, 1.10, and 1.02, i.e. 30/20, 110/100, 510/500, respectively. The condition of the largest fish is very nearly average, while the condition of the smallest fish is quite high. If 10 g extra mass is more beneficial to a small fish than to a large fish, the ratio index of condition may be superior. In reality, there is no way to tell from mass and length alone which index is best. The best index can only be determined by some external criterion, such as whether the index is highly correlated with survival during a food shortage. Furthermore, this index may not be size independent, and the criteria for building the best allometric model for scaling mass to length may not be successful in building an index that is valid for capturing the biology of interest. The critical process of establishing the criterion validity of a condition index is rarely undertaken, consequently, the value of many condition indices is questionable (but see Juliano, 1986 for a nice example of a validated index).

Further evidence for the need to validate condition indices can be gleaned from an analysis of the correlations among alternative condition indices for desert tortoises (Table 1.2; cf. Reist, 1985). In Table 1.2, nine condition indices are calculated, three simple ratios, three regression-dependent ratios, and three residual indices. The correlations among these different indices vary substantially from strong negative correlations ($r = -0.87$) to strong positive correlations ($r = 0.99$) to virtually no correlation ($r = 0.02$). These condition indices are not equivalent. Some of the indices may be good indicators of fat content, others may be better indicators of resistance to infection, yet others may be good indicators of ability to compete for mates, avoid predators, or whatever. Different indices may be needed for different measures, i.e. concepts of condition. Hence, indices must be validated for their intended use, otherwise they are wishful thinking that may be misleading.

If the intended meaning of a condition index is clearly specified, e.g. fat mass, ability to outrun a predator, withstand starvation, produce the most offspring, etc., the best index can be chosen from among a set of alternatives. The best condition index is the one most highly correlated with the biological variable it is intended to describe. For example, a number of studies have found condition indices that are significantly correlated with fat mass (Bailey, 1979; Iverson & Vohs, 1982; Johnson *et al.*, 1985; Brown & Murphy, 1991; Bonnett & Naulleau, 1994; Conway *et al.*, 1994; Hout *et al.*, 1995; Lambert & Dutil, 1997). The question then arises, does validation for one population or species or taxon at any particular

Table 1.2. *Correlation matrix for nine condition indices calculated from the desert tortoise data in Fig. 1.1b*

		Regression residuals			Ratios with division by predicted mass or log mass			Simple ratios		
		Linear	Log	Non-linear	Linear	Log	Non-linear	M/L	M/L^3	$\log M/\log L$
Residuals	Linear	1	0.49	0.54	−0.31	0.42	0.30	0.27	0.44	0.02
	Log		1.00	0.70	−0.59	0.99	0.97	0.27	0.87	0.38
	Non-linear			1.00	−0.14	0.61	0.64	0.25	0.49	0.25
Ratios (predicted)	Linear				1.00	−0.63	−0.57	0.36	−0.87	0.30
	Log					1.00	0.97	0.23	0.89	0.37
	Non-linear						1.00	0.24	0.83	0.43
Simple ratios	M/L							1.00	−0.19	0.93
	M/L^3								1.00	−0.09
	$\log M/\log L$									1.00

Notes:
Residuals are from the linear regression of mass (M) on length (L), the linear regression of log mass on log length, and the non-linear regression of mass on length, respectively. Ratios with division by the predicted value are mass or log mass divided by mass or log mass predicted by the regression.

time generalize to other populations or species or taxa or at other times? It may be possible to construct broadly applicable condition indices for some variables such as body fat percentage for some taxa (Blem, 1990). But it also may not prove possible to formulate indices with broad applicability (cf. Castro & Myers, 1990; Liao *et al.*, 1995; Jonas *et al.*, 1996; Godinho, 1997). More serious efforts are needed to establish whether broadly applicable condition indices exist. If they do not, every study will need to incorporate a validation of any condition index it uses. This result would severely limit the utility of condition indices. Moreover, the need to validate condition indices may make the allometric considerations already discussed irrelevant, because the point is not obtaining size-independent indices; rather it is obtaining valid measures for the concept of interest. These valid measures may not be size independent. Hence, if valid size-dependent indices are found, size may need to be included in any analyses in order to remove, i.e. cancel out its effects and thereby evaluate size-independent condition.

Condition indices as indicators of fat: a special problem

Either explicitly or implicitly, condition indices often are used as indicators of an animal's fat mass or of the fraction of the animal composed of fat. It might appear reasonable to regress mass on length and then to use either the residual from this regression or the ratio mass/predicted mass as an index of condition. Unfortunately, such measures of condition may be poor indicators of fat mass (Weatherhead & Brown, 1996). For example, assume that we regress mass on length and calculate a residual index of condition that is intended to describe the animal's fat mass. The regression of mass on length gives the best prediction for the animal's total mass, which includes both fat and lean mass. The regression is not the best prediction line for an animal's lean mass, so the residual will not reflect the mass of the animal's fat. Likewise, a ratio index of mass divided by predicted mass does not reflect the ratio of total mass (fat plus lean) to lean mass, because the mass predicted by the regression is not lean mass. Because regressions of mass on length (or transforms of length) do not partition lean mass from fat mass, condition indices based on such regressions are likely to be flawed indicators of fat mass. Weatherhead and Brown (1996) showed that residual condition indices formed from regressions of mass on length were substantially worse indicators of fat mass than regressions that accounted for the scaling of lean mass with length in snakes. Their analysis may help explain why residual condition indices had near zero correlations with fat content in mice (Krebs & Singleton, 1993).

Direct analysis of mass and length: an alternative to condition indices

Given that condition indices should be validated but rarely are, we suggest that there often is little value to their use. Despite the potential problems with condition indices, our lack of enthusiasm for them does not mean that mass and length data should not be analysed. Rather, it seems reasonable that morphological data be analysed directly rather than indirectly via a condition index. After all, if a condition index merely summarizes the relationship between mass and length so that it can be correlated with another variable of interest, e.g. starvation resistance, what is the advantage of constructing a condition index? Why not analyse the effects of mass and length data directly?

Instead of constructing an index whose value is not established, a better alternative to formulating condition indices is to analyse the available data directly. This direct approach is better biologically because unsubstantiated arguments about condition are not needed. To contrast the two approaches, suppose you measure territory size, mass, snout–vent length, age, and locomotor performance in a population of lizards. Advocates of condition indices might use mass and length to construct one of the numerous condition indices in widespread use and then regress territory size on condition index, age, and locomotor performance to assess the effects of condition on territory size. Again, a major problem with this approach is that it is unclear what condition means.

As an alternative to the construction of this nebulous condition index, we suggest analysing the data directly. For example, regress territory size on mass, length, age, and locomotor performance and include an interaction term (mass by length) if appropriate. Standard diagnostic procedures may be used to fit an appropriate model. If the relationship is non-linear then non-linear regression may be used instead of linear regression or perhaps logarithmic or other transformation of some of the variables will be beneficial. In a direct analysis that omits the condition index, the effects of mass and length and so on are straightforward to assess. Direct analyses of the data eliminate any need for special pleading about the informative value of an ill-defined condition index.

If a biologist is convinced that the relationship between mass and length contains critical information, direct analyses of the data are still possible. For example, you might fit the model $y = \delta + \alpha(\text{mass} - \text{length}^{\beta}) + e$. We constructed an empirical model like this and compared it with the results of analyses that used a variety of residual-based indices. We analysed data on mating success, mass, and snout–vent length of water snakes (*Nerodia sipedon*). These data were kindly provided by P. Weatherhead. First, we fitted the direct model: $(\text{mating success})^{1/2} = \delta + \alpha\ (\text{mass} - (\text{snout–vent length})^{\beta}) + e$. We then compared the fit of this model to the fits of the regressions of square root of mating success on residuals from (i) the regression of mass on length, (ii) the regression of mass on length^3, (iii) the regression of $\log_e$ mass on $\log_e$ length, (iv) the non-linear regression of mass on length with no intercept, and (v) the non-linear regression of mass on length with an intercept. The direct analysis model fits better than any of the residual models (Table 1.3). In a sense, the direct model defines a new empirically fitted condition index $(\text{mass} - (\text{snout–vent length})^{\beta})$. The advantage to the term is that, by construction, it is the best fit to the response variable. If the regression coefficient α on $(\text{mass} - (\text{snout–vent length})^{\beta})$ is statistically significant, the term, or condition if you prefer, is empirically validated in that sample. Nonetheless, we caution that, while the term might be a good indicator of mating success, it might not be a good indicator of any other response, i.e. it might not be indicative of ability to resist infection or avoid predation or survive food shortage. In any case, the point that we wish to make is that, in the absence of external validation, a direct analysis of the data is more appropriate than constructing unvalidated condition indices. The direct analysis essentially tests the validity of the morphometric data as indicators of the response variable of interest.

Besides the concerns already raised, there is one last reason for avoiding the analysis of condition indices. This concern applies only to indices derived from regressions, e.g. the ratio of mass to predicted mass or the residual mass or other statistically fitted models. The use of residuals and predicted values in subsequent regression analysis may lead to erroneous statistical, and hence biological, inferences (cf. Morton *et al.*, 1991; Smith & McAleer, 1994). In general, residuals and predicted values should not be treated as raw data. Regressions that include previously generated residual or predicted values are potentially seriously compromised (Murphy & Topel, 1985). This topic is one that needs further investigation, but it is potentially important because it is not uncommon for biologists to use residuals and predicted values in subsequent regressions.

Table 1.3. *Comparison of direct fitted model with residual-based condition indices for the water snake*

Residual condition index (CI)	R^2 for preliminary regression	Model	R^2 for model
None	None	$y = 1.76 + 0.0219\,(\text{mass} - \text{SVL}^{1.23}) + e$	0.261
$\text{Mass} = -145 + 4.78\ \text{SVL} + \text{CI}$	0.825	$y = 1.23 + 0.0211\,(\text{CI}) + e$	0.145
$\text{Mass} = 20.8 + 0.000562\ \text{SVL}^{3} + \text{CI}$	0.804	$y = 1.23 + 0.0221\,(\text{CI}) + e$	0.179
$\text{Ln(mass)} = -6.06 + 2.70\ \ln(\text{SVL}) + \text{CI}$	0.871	$y = 1.23 + 2.67\,(\text{CI}) + e$	0.212
$\text{Mass} = 0.00895\ \text{SVL}^{2.36} + \text{CI}$	0.821	$y = 1.23 + 0.0238\,(\text{CI}) + e$	0.189
$\text{Mass} = -55.8 + 0.301\ \text{SVL}^{1.59} + \text{CI}$	0.831	$y = 1.23 + 0.0224\,(\text{CI}) + e$	0.159

In conclusion, in the absence of external verification, the construction of condition indices based on the relationships between external morphological variables is unjustified. Instead of constructing condition indices, we advocate analysing data directly. If a clearly specified metric for condition, e.g. fat mass, resistance to infection, ability to win agonistic encounters is of interest, the metric of condition should be measured or else a verified procedure for predicting condition should be developed.

Recommendations

(i) We recommend that biologists only use the term condition index when they clearly articulate its intended meaning. Nebulously defined condition indices are of little value.
(ii) Condition indices should be validated. Without validation, the construction of ratios or residuals to describe condition is of little value.
(iii) If condition indices are not validated, they are more wishful thinking than worthwhile biology. A better alternative to the formulation of superfluous indices is a direct analysis of the available data.

Acknowledgements

We thank Todd Escue, Bobby Espinoza, Alan Gubanich, Kenneth Geluso, Mary Peacock, Richard Rust, and Patrick Weatherhead for allowing us to use their original data on desert tortoises, chuckwallas, orange-crowned warblers, white-ankled mice, cutthroat trout, mason bees, and water snakes, respectively. We thank Patrick Weatherhead, Heather Goulding, and two anonymous reviewers for comments on the manuscript. We thank Mike McAleer for responding to numerous e-mails about the use of generated regressors and residuals in subsequent analyses.

REFERENCES

Albrecht, G.H., Gelvin, B.R. & Hartman, S.E. (1993). Ratios as a size adjustment in morphometrics. *American Journal of Physical Anthropology*, **91**, 441–68.

Allison, D.B., Paultre, F., Goran, M.I., Poehlman, E.T. & Heymsfield, S.B. (1995). Statistical considerations regarding the use of ratios to adjust data. *International Journal of Obesity*, **19**, 644–52.

Atchley, W.R., Gaskins, C.T. & Anderson, D. (1976). Statistical properties of ratios. I. Empirical results. *Systematic Zoology*, **25**, 137–48.

Bailey, R.O. (1979). Methods of estimating total lipid content in the redhead duck

(*Aythya americana*) and an evaluation of condition indices. *Canadian Journal of Zoology*, **57**, 1830–3.

Batzli, G.O. & Esseks, E. (1992). Body fat as an indicator of nutritional condition for the brown lemming. *Journal of Mammalogy*, **73**, 431–9.

Beaupre, S.J. & Dunham, A.E. (1995). A comparison of ratio-based and covariance analyses of a nutritional data set. *Functional Ecology*, **9**, 876–80.

Blem, C.R. (1990). Avian energy storage. *Current Ornithology*, **7**, 59–113.

Bolger, T. & Connolly, P.L. (1989). The selection of suitable indices for the measurement and analysis of fish condition. *Journal of Fish Biology*, **34**, 171–82.

Bollen, K.A. (1989). *Structural Equations with Latent Variables.* New York: John Wiley.

Bollen, K.A. & Ward, S. (1979). Ratio variables in aggregate data analysis: their uses, problems, and alternatives. *Sociological Methods and Research*, **7**, 431–50.

Bonnett, X. & Naulleau, G. (1994). A body condition index (BCI) in snakes to study reproduction. *Comptes Rendus de L'Academie des Sciences Serie III- Sciences de la Vie-Life Science*, **317**, 34–41.

Bookstein, F.L. (1989). 'Size and shape': a comment on semantics. *Systematic Zoology*, **38**, 173–80.

Brown, M.E. (1996). Assessing body condition in birds. *Current Ornithology*, **13**, 67–135.

Brown, M.L. & Murphy, B.R. (1991). Relationships of relative weight (Wr) to proximate composition of juvenile striped bass and hybrid striped bass. *Transactions of the American Fisheries Society*, **120**, 509–18.

Castro, G. & Myers, J.P. (1990). Validity of predictive equations for total body fat for sanderlings from different nonbreeding areas. *Condor*, **92**, 205–9.

Cattet, M. (1990). Predicting nutritional condition in black bears and polar bears on the basis of morphological and physiological measurements. *Canadian Journal of Zoology*, **68**, 32–9.

Cone, R.S. (1989). The need to reconsider the use of condition indices in fishery science. *Transactions of the American Fisheries Society*, **118**, 510–14.

Conway, C.J., Eddleman, W.R. & Simpson, K.L. (1994). Evaluation of lipid indices of the wood thrush. *Condor*, **96**, 783–90.

Draper, N.R. & Smith, H. (1981). *Applied Regression Analysis.* New York: John Wiley.

Droney, D.C. (1998). The influence of the nutritional content of the adult male diet on testis mass, body condition and courtship vigour in an Hawaiian *Drosophila*. *Functional Ecology*, **12**, 920–8.

Dunn, J.P. & Chapman, J.A. (1983). Reproduction, physiological responses, age structure, and food habits of raccoon in Maryland, USA. *Zeitschrift für Säugertierkunde*, **48**, 161–75.

Framstad, E.S., Engen, S. & Stenseth, N.C. (1985). Regression analysis, residual analysis, and missing variables in regression models. *Oikos*, **44**, 319–23.

Freeman, S. & Jackson, W.M. (1990). Univariate metrics are not adequate to measure avian body size. *Auk*, **107**, 69–74.

Fuller, W.A. (1987). *Measurement Error Models.* New York: John Wiley.

Fulton, T.W. (1904). The rate of growth of fishes. *Annual Report of the Fishery Board for Scotland*, **22**(Part III),141–241.

Gingerich, P.D. (1995). Statistical power of EDF tests of normality and the sample size required to distinguish geometric-normal (lognormal) from arithmetic-normal distributions of low variability. *Journal of Theoretical Biology*, **173**, 125–36.

Godinho, A.L. (1997). Weight-length relationship and condition of the characiform *Triportheus guentheri*. *Environmental Biology of Fishes*, **50**, 319–30.

Harvey, P.H. & Pagel, M.D. (1991). *The Comparative Method in Evolutionary Biology*. Oxford: Oxford University Press.

Hayes, D.B., Brodziak, J.K.T. & O'Gorman, J.B. (1995). Efficiency and bias of estimators and sampling designs for determining length weight relationships of fish. *Canadian Journal of Fisheries and Aquatic Sciences*, **52**, 84–92.

Hayes, J.P. & Shonkwiler, J.S. (1996). Analyzing mass-independent data. *Physiological Zoology*, **69**, 974–80.

Hogg, R.V. & Craig, A.T. (1995). *Introduction to Mathematical Statistics*. Upper Saddle River, New Jersey: Prentice Hall.

Hood, L.C., Boersma, P.D. & Wingfield, J.C. (1998). The adrenocortical response to stress in incubating Magellanic penguins (*Spheniscus magellanicus*). *Auk*, **115**, 76–84.

Hout, J., Poulle, M.-L. & Crête, M. (1995). Evaluation of several condition indices for assessment of coyote (*Canis latrans*) body composition. *Canadian Journal of Zoology*, **73**, 1620–4.

Iverson, G.C. & Vohs, P.A., Jr. (1982). Estimating lipid content of sandhill cranes from anatomical measurements. *Journal of Wildlife Management*, **46**, 478–83.

Jakob, E.M., Marshall, S.D. & Uetz, G.W. (1996). Estimating fitness: a comparison of body condition indices. *Oikos*, **77**, 61–7.

Jasienski, M. & Bazzaz, F. A. (1999). The fallacy of ratios and the testability of models in biology. *Oikos*, **84**, 321–6.

Johnsen, T.S., Hengeveld, J.D., Blank, J.L., Yasukawa, K. & Nolan, V., Jr. (1996). Epaulet brightness and condition in female red-winged blackbirds. *Auk*, **113**, 356–62.

Johnson, D.H., Krapu, G.L., Reinicke, K.J. & Jorde, D.G. (1985). An evaluation of condition indices for birds. *Journal of Wildlife Management*, **49**, 569–75.

Jolicoeur, P. (1975). Linear regressions in fishery research: some comments. *Journal of the Fisheries Research Board of Canada*, **32**, 1491–4.

Jonas, J.L., Kraft, C.E. & Margenau, T.L. (1996). Assessment of seasonal energy density and condition in age-0 and age-1 muskellunge. *Transactions of the American Fisheries Society*, **125**, 203–10.

Jorgensen, C.B. (1983). Pattern of growth in a temperate zone anuran (*Bufo viridis* Laur.). *Journal of Experimental Zoology*, **227**, 433–9.

Juliano, S.A. (1986). Food limitation of reproduction and survival for populations of *Brachinus* (Coleoptera: Carabidae). *Ecology*, **67**, 1036–45.

Jungers, W.L., Falsetti, A.B. & Wall, C.E. (1995). Shape, relative size, and size-adjustments in morphometrics. *Yearbook of Physical Anthropology*, **38**, 137–61.

Krebs, C.J. & Singleton, G.R. (1993). Indices of condition for small mammals. *Australian Journal of Zoology*, **41**, 317–23.

Kronmal, R.A. (1993). Spurious correlations and the fallacy of the ratio standard revisited. *Journal of the Royal Statistical Society A*, **156**, 379–92.

Kronmal, R.A. (1995). Author's reply. *Journal of the Royal Statistical Society A*, **158**, 623–5.

LaBarbera, M. (1989). Analyzing body size as a factor in ecology and evolution. *Annual Review of Ecology and Systematics*, **20**, 97–117.

Lambert, Y. & Dutil, J-D. (1997). Can simple condition indices be used to monitor and quantify seasonal changes in energy reserves of Atlantic cod (*Gadus morhua*)? *Canadian Journal of Fisheries and Aquatic Sciences*, **54** (Suppl. 1), 104–12.

Le Cren, E.D. (1951). The length-weight relationship and seasonal cycle in gonad weight and condition in the perch (*Perca fluviatilis*). *Journal of Animal Ecology*, **20**, 201–19.

Liao, H., Pierce, C.L., Wahl, D.H., Rasmussen, J.B. & Leggett, W.C. (1995). Relative weight as a field assessment tool: relationships with growth, prey biomass, and environmental conditions. *Transactions of the American Fisheries Society*, **124**, 387–400.

Manly, B.F.J. (1986). *Multivariate Statistical Methods: A Primer*. London: Chapman and Hall.

McArdle, B.H. (1988). The structural relationship – regression in biology. *Canadian Journal of Zoology*, **66**, 2329–39.

Mood, A.M., Graybill, F.A. & Boes, D.C. (1974). *Introduction to the Theory of Statistics*. New York: McGraw-Hill.

Morton, J.M., Kirkpatrick, R.L. & Smith, E.P. (1991). Comments on estimating total body lipids from measures of lean mass. *Condor*, **93**, 463–5.

Murphy, K.M. & Topel, R.H. (1985). Estimation and inference in two-step econometric models. *Journal of Business and Economic Statistics*, **3**, 370–9.

Neville, A.M. & Holder, R.L. (1995). Letter to the editors. Spurious correlations and the fallacy of the ratio standard revisited. *Journal of the Royal Statistical Society A*, **158**, 619–21.

Packard, G.C. & Boardman, T.J. (1999). The use of percentage and size-specific indices to normalize physiological data for variation in body size: wasted time, wasted effort? *Comparative Biochemistry and Physiology A*, **122**, 37–44.

Pearson, K. (1897). Mathematical contributions to the theory of evolution. On a form of spurious correlation which may arise when indices are used in the measurement of organs. *Proceedings of the Royal Society of London*, **60**, 489–98.

Piersma, T. & Davidson, N.C. (1991). Confusions of mass and size. *Auk*, **108**, 441–4.

Rayner, J.M.V. (1985). Linear relations in biomechanics: the statistics of scaling functions. *Journal of Zoology*, **206**, 415–39.

Reist, J.D. (1985). An empirical evaluation of several univariate methods that adjust for size variation in morphometric data. *Canadian Journal of Zoology*, **63**, 1429–39.

Ricker, W.E. (1973). Linear regressions in fishery research. *Journal of the Fisheries Research Board of Canada*, **30**, 409–34.

Ricker, W.E. (1975*a*). A note concerning Professor Jolicoeur's comments. *Journal of the Fisheries Research Board of Canada*, **32**, 1494–8.

Ricker, W.E. (1975*b*). Computation and interpretation of the biological statistics of fish populations. *Bulletin of the Fisheries Research Board of Canada*, **191**, 1–382.

Rising, J.D. & Somers, K.M. (1989). The measurement of overall size in birds. *Auk*, **106**, 666–74.

Royle, N.J. & Hamer, K.C. (1998) Hatching asynchrony and sibling size hierarchies in gulls: effects on parental investment decisions, brood reduction and reproductive success. *Journal of Avian Biology*, **29**, 266–72.

Smith, R.J. (1993). Logarithmic transformation bias in allometry. *American Journal of Physical Anthropology*, **90**, 215–28.

Smith, J. & McAleer, M. (1994). Newey–West covariance matrix estimates for models with generated regressors. *Applied Economics*, **26**, 635–40.

Tuomi, J. & Jormalainen, V. (1988). Components of reproductive effort in the aquatic isopod *Idotea baltica*. *Oikos*, **52**, 250–4.

van der Meer, J. & Piersma, T. (1994). Physiologically inspired regression models for

estimating and predicting nutrient stores and their composition in birds. *Physiological Zoology*, **67**, 305–29.
van Marken Lichtenbelt, W.D., Wesselingh, R.A., Vogel, J.T. & Albers, K.B.M. (1993). Energy budgets in free-living green iguanas in a seasonal environment. *Ecology*, **74**, 1157–72.
Viggers, K.L., Lindenmayer, D.B., Cunningham, R.B. & Donnelly, C.F. (1998). Estimating body condition in the mountain brushtail possum, *Trichosurus caninus*. *Wildlife Research*, **25**, 499–509.
Weatherhead, P.J. & Brown, G.P. (1996). Measurement versus estimation of condition in snakes. *Canadian Journal of Zoology*, **74**, 1617–21.
Wiklund, C.G. (1996). Body length and wing length provide univariate estimates of overall body size in the Merlin. *Condor*, **98**, 581–8.
Zar, J.H. (1968). Calculation and miscalculation of the allometric equation as a model in biological data. *BioScience*, **18**, 1118–20.

D. SCOTT REYNOLDS AND THOMAS H. KUNZ

2

Standard methods for destructive body composition analysis

Introduction

Body composition analysis can provide insight into many aspects of an animal's physiology, ecology, and life history. Historically, this analysis has required sacrificing individuals to obtain accurate estimates of body composition. Recent advances in technology and innovative approaches towards developing non-destructive relationships have begun to advance our understanding of body composition without sacrificing individuals. However, these new approaches must still be validated for both accuracy and precision using direct measurements. The purpose of this chapter is to present standard methods for body composition analysis, point out some of the problems encountered when attempting to analyse body composition, and provide guidance for avoiding some of these problems.

Statistical issues when measuring body composition

Body composition is a dynamic variable that often shows a high level of variation. When it has been studied, variation in body composition has been found at the level of the individual, population, and species. Although body composition analysis has been used at these three levels, most studies have focused on differences in body composition within a population. Sources of this variation can be both intrinsic (such as age, sex, or reproductive state) and extrinsic (such as climate or resource availability). In addition to natural variation, techniques used to study body composition may also introduce measurement error. Variation, irrespective of the source, must be controlled as much as possible in studies of body composition.

Regardless of the taxa being investigated, several steps should be taken to account for natural variation and minimize experimental error.

Sample size

When planning an investigation, the question of an appropriate sample size is crucial. When destructive techniques are used, sample sizes are often reduced to minimize the number of individuals sacrificed. For example, many studies have estimated the body composition of a species using fewer than 15 individuals (Worthy & Lavigne, 1983; Fiorotto *et al.*, 1987; Piersma, 1988; Bachman, 1994; T.H. Kunz and A.L. Stern, unpublished data). In some studies that have used males and females or adult and young, larger numbers of individuals are often reduced to statistical groups of less than ten individuals (Ewing *et al.*, 1970; O'Farrell & Studier, 1973; Millar, 1975). One consideration that must be given to sample size is the level of variation in the population. The focus of many investigations is to evaluate a difference between two populations (statistical or ecological) in some character. However, the ability to detect this difference depends on the level of variation that exists between individuals within each population, and the degree of variation (error) that is produced by the research methodology. There are several precautions that a researcher can undertake to control these sources of variation.

Effect-size analysis

Effect-size analysis determines a minimum sample size that is needed to detect a difference between two sample populations given an estimate of both between-population (effect size) and within-population (population error variance) variation. Statistical and experimental factors can influence the ability to detect a difference between two populations. For a more detailed review of effect size analysis, see Cohen (1977). Statistical factors, such as level of significance (Type I error rate) and power of the test (Type II error rate) can be chosen *a priori*. However, experimental factors depend on the level of variation in the population (population error variance σ^2) and the effect size (ES) being investigated.

Level of significance (α)

The convention for acceptable Type I experimentwise error rate is $\alpha = 0.05$.

Power ($1 - \beta$)

The Type II error rate (β) is a function of α and the difference in means between the null and alternative hypothesis. Large differences in the mean of the alternative hypothesis (H_a), or small standard error rates in their sampling distribution, produce smaller β, and consequently a more powerful test. Increasing sample size reduces the standard error and therefore increases the power of a test.

Population error variance (σ^2)

The population error variance is seldom known in scientific research. However, a sample variance (s^2) can be obtained from preliminary data collection and used to determine a standardized effect size.

Effect Size (ES)

Effect size is the desired difference to be detected between H_0 and H_a. The larger the difference between the two samples, the easier it is to detect a significant difference.

Standardized effect size

$$d = \frac{ES}{\sigma}$$

Determining minimum sample size using the Z statistic (for details on the Z statistic, see Sokal & Rohlf, 1982):

For one-tailed tests:

$$n = \frac{(z_\beta - z_\alpha)^2}{d^2}$$

For two-tailed tests:

$$n = \frac{(z_\beta - z_{\alpha/2})^2}{d^2}$$

Within-individual variation

The most common aspect of within-individual variation of interest to ecologists is how body composition changes as an individual grows, matures and reproduces. These considerations are the main disadvantages of destructive sampling. However, sampling selective tissues or organs rather than using whole-body analysis is another aspect of within-individual variation that is often ignored in studies on large animals, especially if one incorrectly assumes that body composition is a homogeneous property of an individual. Although whole-body composition analysis is more tedious for large animals, techniques exist for reliable subsampling that produce accurate whole-body composition estimates (see preparing samples for analysis).

Controlling experimental error

Two types of experimental error are important to identify and control in body composition analysis. Conducting preliminary research that establishes a protocol to control both sample collection and variation in sample preparation can minimize within-experiment error. The second type (between-experiment error) could conceivably be controlled or minimized by using published techniques to analyse samples. However, materials and methods sections of published papers seldom contain sufficient detail that would allow researchers to replicate experiments. Moreover, no standardized methods currently exist for body composition analysis. Although these obstacles cannot be avoided in retrospective studies, future papers should be published in sufficient detail so that other researchers can replicate the methods employed.

Sample collection and preparation

A flow diagram of stepwise procedures for conducting body composition analysis is shown in Fig. 2.1.

Capturing specimens

Daily and seasonal variation in time of capture can markedly affect estimates of body composition for individuals. Although seasonal variation in body composition is well documented, daily variation in total body mass and total body water is common, and research has shown that body fat can also exhibit daily variation (Meijer *et al.*, 1994; Nestlet *et al.*, 1996). Controlling for daily and seasonal variation in body composition is particularly important for small homeothermic species (Licht & Leitner, 1967; McNab, 1976; Tidemann, 1982; Kunz *et al.*, 1998). When not the focus of a study, daily and seasonal variation can be minimized by collecting individuals at the same time each day and season, respectively.

Age- and sex-dependent differences in body composition are also common; the latter effect is most prevalent among sexual dimorphic species. As part of any investigation that contains both males and females, sex-specific body composition should be tested before combining data. A similar precaution is warranted when studies include individuals of varying age classes. At a minimum, we suggest for species that have determinate growth that separate comparisons be made for juveniles and adults. In species that exhibit indeterminant growth, the effect of increasing body size with age is less clear, and whether mass-independent variables completely remove this effect is subject to debate. Regardless of the procedure used, different classes of individuals should not be pooled until it has been shown that they are not statistically different.

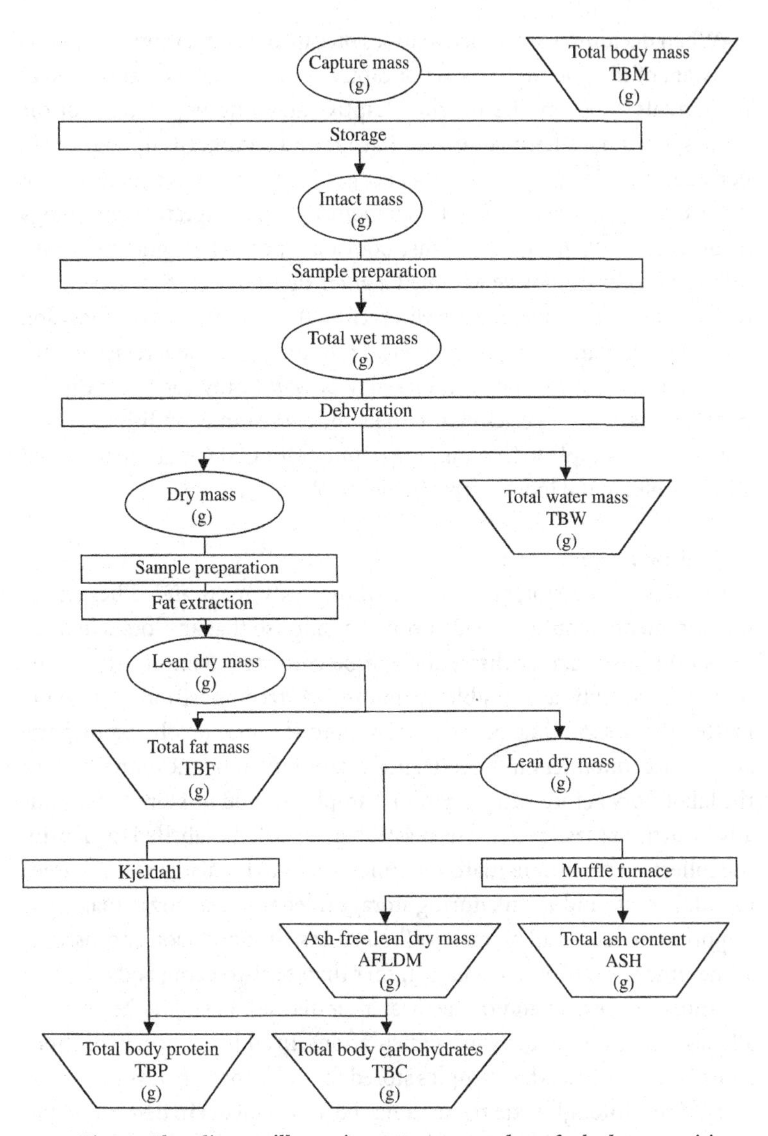

Fig. 2.1. Flow diagram illustrating stepwise procedures for body composition analysis. Methods are depicted by rectangular windows, empirical values are shown as oval windows, and derived values are indicated by keystone-shaped windows.

When deciding to conduct analyses on captive populations, one must be aware of the potential impact of captivity on the body composition of individuals. Most studies in which captive and wild populations of the same species have been compared have found significant differences in body composition (Hayward, 1965; Ledger, 1968; Millar, 1975, 1981), most likely due to differences in diet, metabolic rate, and digestive physiology (Ledger, 1968; Webb, 1992). Because body fat tends to be higher in captive populations, it cannot be assumed *a priori* that relationships developed for non-destructive techniques will accurately predict body composition in both captive and wild populations. Thus, unless verified, relationships developed for captive populations may be valid only for other captive populations of that species maintained under the same conditions. These same concerns apply to free-ranging populations that have been provided with supplemental food sources (Arnold & Ankney, 1997).

Storing specimens

Regardless of the storage method, whole body mass and other intact measurements should be made prior to storage so that any losses in body mass due to storage or dissection can be corrected. Samples should be frozen as quickly as possible to minimize decomposition of organic matter. When sampling locations are remote, this may involve using portable dry ice containers at the collection site or returning the individuals to the laboratory before euthanization. Samples should be stored individually in airtight plastic bags. The plastic bags should be labelled to identify the collection conditions (date and time) and individual identity. Samples may dehydrate and shrink during storage (Fleharty *et al.*, 1973), making it important to record all biometrics (body mass, limb and foot lengths, etc.) at the time of capture. This mass, rather than post-freezing body mass or the sum of the organ and tissue mass after dissection, should be used for all mass-specific total body mass calculations. In addition to dehydration, Paine (1971) suggests that samples stored for more than 30 days can begin to oxidize, although little research has been conducted to determine the magnitude of this effect. Other researchers have suggested that samples can be stored for up to 6 months with no appreciable effect on body composition (Pierson & Stack, 1988).

Preparing samples for analysis

Different taxa may require alternative preparations for body composition analysis. One of the greatest potential sources of error in sample preparation comes from differences in the removal of selected tissues or organs, including removal of stomach contents and reproductive tissue prior to analysis. Although most studies of body composition remove the stomach contents (and crop contents of birds) prior to analysis, whether this material should be removed prior to analysis often depends on the focus of the study. For example, energetic studies often remove stomach contents because they represent unassimilated energy that an individual would not be capable of utilizing. However, ecological studies, particularly those that evaluate energy flow through a community, should view the individual the same way a predator views its prey. This would mean retaining stomach contents but removing culled items such as appendages or feathers.

The rationale for removing materials is seldom explained or justified. For example, Jenni-Eiermann (1989) removed the viscera, liver, heart, stomach, head and feathers prior to analysis, whereas Thompson and Drobney (1996) removed the digestive tract (including the liver and pancreas), left breast muscle, left leg muscle, and heart; both of these studies involved the same genera of bird (*Aythya* spp.). Other examples of this variation provide a clear illustration of the potential impact of sample preparation on the resulting body composition estimates. In three separate studies of dabbling geese (*Anas* spp.), Reinecke and Stone (1982) removed only the feathers, Wishart (1979) removed the stomach contents, and Ringelman and Szymczak (1985) removed the bills and the feet. These three studies were destined to produce different estimates of body composition owing to variation in protocols for preparation irrespective of differences in whole-body composition. The point of highlighting these studies is to highlight two issues that are prevalent in body composition analyses; the lack of standardized protocols and lack of explanations or rationale that would allow researchers to compare the impact of different techniques on the results obtained.

When researchers justify the removal of additional materials, it is most often done to conduct different analyses on the same individuals for related studies. For example, this may include the removal of materials to develop condition indices from bone marrow (Winstanley *et al.*, 1998), liver (Mascher & Marcstrom, 1976), muscle (Brittas & Marcstrom, 1982), or fat pads (Lynch, 1973; Morton, 1975). In our opinion, whole body composition analysis should include all components of the organism unless the removal of each component is adequately justified and the qualitative impacts of that removal are investigated.

Once the carcass has been obtained, the level of additional preparation depends on the size of the animal and the analysis being performed. Specifically, the mass of the sample being analysed should be appropriate for each of the techniques being employed. For small animals (body mass <1.0 g), this may require pooling individuals. For small organisms, precision and detection limitations of equipment may require pooling carcasses from several individuals within a single sample, thus obscuring within-individual variation. While whole-body composition analysis is generally straightforward in medium-sized organisms (1.0 g < body mass >1.0 kg), it is often not feasible for large organisms (>1.0 kg). For large organisms, researchers should homogenize the sample by processing the whole carcass in a grinder, using successively finer mesh sizes (see Sample preparation) and then taking a mass of the homogenate for body composition analysis. Studies that subsample a non-homogenized carcass or use a single tissue type without attempting to correlate these samples with whole-body composition may not accurately represent the whole-body composition, but rather produce estimates that are only valid for a specific tissue or subsample.

Whole-body composition analysis

This chapter summarizes the most common methods used to measure five major components of body composition; water, fat, protein, carbohydrates, and inorganic constituents. In practice, several of these components are often later pooled. For example, two-compartment models separate the body composition into fat and lean mass (water, protein, carbohydrates, and inorganic constituents), whereas three-compartment models separate fat, water, and lean dry mass (protein, carbohydrates, and inorganic constituents). In addition to variation in how these components are measured, there is considerable variation in how the components are recorded.

Body composition analysis directly measures the total mass of each component. However, because there is variation between individuals in total body mass, differences in these mass-dependent measures may not reflect differences in relative body composition. Thus, mass-independent variables are often used to control for the effects of variation in body mass between individuals. How these mass-dependent variables are used in statistical tests has been the subject of considerable debate (Angervall & Carltrom, 1963; Miller & Weil, 1963; Atchley *et al.*, 1976).

Water mass

Water mass is the predominant component of all living organisms, comprising from 62–74% of body mass in a variety of birds and small mammals (Fleharty *et al.*, 1973; Holmes, 1976), although this value is often less in larger animals (Frisch, 1987; Arnould *et al.*, 1996). Water mass of newborns and young animals is higher than adult animals, often exceeding 80% of total body mass (Widdowson, 1950; Myrcha & Walkowa, 1968; Lochmiller *et al.*, 1983). Lastly, water mass is known to fluctuate seasonally (Fleharty *et al.*, 1973; Schreiber & Johnson, 1975; Kunz *et al.*, 1998), as well as with reproductive condition (Galster & Morrison, 1976; Kiell & Millar, 1980) in a variety of smaller animals.

An accurate measurement of water mass is an important component of any study attempting to assess body composition. This is particularly true for non-destructive techniques, such as isotope dilution, that rely on estimates of total body water to predict other body compartments (Chapters 3, 5 and 6, this volume). Water mass is generally measured as mass-independent total body water (TBW, g water) or mass-dependent water content (WC (g water/g body mass)) or water index (WI (g water/g lean dry mass)). Total body water is calculated by subtracting the carcass mass following complete dehydration from the original capture mass; the difference being the total mass of water evaporated during dehydration.

Dehydration is generally achieved by one of three methods: freeze-drying, vacuum ovens, or convection ovens. Freeze-drying is often considered the best method because it does not change tissue composition (Kerr *et al.*, 1982; Pierson & Stack, 1988). However, the magnitude of error that is produced using convection or vacuum ovens can be quite low (Kerr *et al.*, 1982). Evaluation of these different methods is also hampered by a general lack of disclosure of drying temperatures in published studies (Kerr *et al.*, 1982). Two primary concerns with oven dehydration are volatilization of organic matter (which increases with oven temperature) and decomposition of the sample (which increases with low oven temperature due to increased drying time). The balance of these trade-offs can shift towards increasing risk of decomposition when the fat content of the samples is high. Because vacuum ovens have lower drying times and temperatures than convection ovens, there is a lower risk of tissue samples being volatized or degraded. However, convection ovens operated between 60 °C and 90 °C appear to offer a reasonable balance between these two constraints.

Dehydration occurs most rapidly and uniformly when the surface area of the sample is increased. For small animals, this can be achieved simply by cutting the carcass into small pieces. Larger animals may require being ground in a commercial meat grinder to generate a homogenate that can be subsampled and used for all subsequent analyses. Variation in water content between individuals can produce variation in the time required to complete dehydration. Thus, frequent measurements of the sample mass should be made to determine when a constant mass has been achieved. Although some studies have suggested dehydration be conducted for a fixed time interval (Gyug & Millar, 1980; Rickart, 1982; Ringelman & Szymczak, 1985), this approach is only advised when the sample has reached a constant mass prior to measuring the total water loss.

Fat mass

Fat represents the major energy storage compartment of animals, and is therefore often the focus of body composition studies. However, because fat mass is also the most variable aspect of an individual's body composition, it is also the most difficult component to estimate. The lower fat-storage capacity and a higher mass-specific metabolic rate (Schmidt-Nielsen, 1994) of smaller animals produces high levels of daily and seasonal variation in fat mass. However, large animals can also exhibit seasonal variation in body fat resulting from reproduction, migration, or hibernation (Weber & Thompson, 1998; Winstanley *et al.*, 1999).

Direct estimates of body fat require the extraction of lipids from the carcass using a variety of polar and non-polar solvents. In general, the preferred solvent is petroleum ether and diethyl ether, used alone or in a mixture with ethyl alcohol; 3:1 mixture of 95% ethyl alcohol and petroleum ether is commonly used in fat extractions. Although chloroform has been used, either alone or with methanol, it is carcinogenic and may extract non-lipids with the fat (Dobush *et al.*, 1985; Pierson & Stack, 1988). Several studies have extracted fat by placing the samples in a sealed, tared pouch submerged in a solvent bath under adequate ventilation (Bligh & Dyer, 1959; Holmes, 1976; Fiorroto *et al.*, 1987; Bachman, 1994). However, most fat extractions are currently performed using a Soxhlet apparatus, although similar devices such as the Goldfisch or Folch apparatus also have been used.

The details of the Soxhlet procedure are described in Sawicka-Kapusta (1975). The Soxhlet apparatus contains four major components; a solvent flask, extractor vessel, a hot plate or other heating element, and a water-cooled condenser (see Fig. 2.2). The samples are placed into a porous extraction thimble within the extractor. The solvent within the flask is

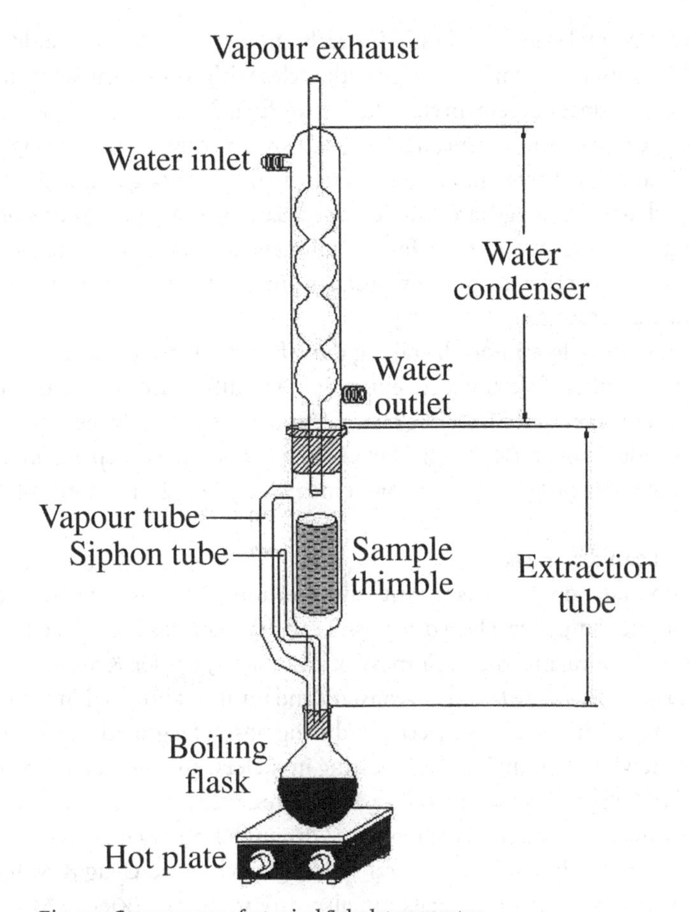

Fig. 2.2. Components of a typical Sohxlet apparatus.

heated with the hot plate; solvent vapours are condensed within the condenser coils and slowly titrate into the extractor flask containing the samples. When the solvent level within the extractor flask reaches the height of the overflow tube within the flask, the solvent is pulled through the tube back into the solvent flask, whereupon it is reheated and the cycle is repeated. The cycle time of this procedure varies with the size of the solvent flask and extractor. The number of cycles required to complete fat extraction depends upon the fat content of the sample, but 25 cycles are usually sufficient to ensure complete extraction. Because the extracted fat is returned to the solvent flask after each cycle, it is recommended that the solvent be replaced periodically for each sample. The solvent should be replaced after the tenth and twentieth cycles with fresh, clear solvent. If the solvent does not remain clear during the final five cycles, it may be necessary to replace the solvent an additional time and continue extracting fat until the solvent remains clear. If protein analysis is not going to be performed, or if a homogenate sample is used, Sawicka-Kapusta (1975) recommends soaking the sample for 30 minutes in 95% ethanol prior to extraction in order to prevent proteins from washing out during the extraction process.

Fat mass is determined by taking the difference between the dry mass of the sample before and after extraction, the difference being the total grams of fat extracted from the carcass. Fat mass is generally measured as mass-independent total body fat (TBF, g fat) or mass-dependent fat content (FC (g fat/g body mass)) or fat index (FI (g fat/g lean dry mass)).

Lean dry mass

The non-fat, non-aqueous constituent of an animal is most often studied as a single component; lean dry mass. Lean dry mass has been referred to as the structural mass or 'basic mass' of an animal (Jameson & Mead, 1964) based on the premise that it is a constant and invariant proportion of total body mass. This is clearly inaccurate during postnatal growth and development when an individual is investing in skeletal and muscular growth (Millar, 1981), during mobilization of body reserves during starvation, or deposition during periods of recovery. Although lean dry mass is known to be constant in adults of several species (Lynch, 1973; Gyug & Millar, 1980), lean mass of individuals can also vary seasonally (Kiell & Millar, 1980; Atkinson *et al.*, 1996; Reynolds & Kunz, 2000). In a three-compartment model, lean dry mass is determined by subtracting total body water and total body fat from the original body mass.

$$\text{LDM} = \text{Body mass} - (\text{TBW} + \text{TBF})$$

Although mass-independent measures of lean dry mass are seldom presented, a lean mass index can be useful (LMI (g LDM/g total body mass)). Often, lean dry mass is partitioned into three base components: proteins, ash, and carbohydrates.

Ash content

The ash component of an organism represents its inorganic mineral content. The two most abundant minerals for vertebrate animals are calcium and phosphorus, the two main elements in bone (Robbins, 1993). Ash content is often measured as the residue remaining after combustion of the lean dry samples in either a muffle furnace or bomb calorimeter. When the two techniques have been compared, the muffle furnace provides a more accurate measure of ash content (Paine, 1971). Bomb calorimetry appears to underestimate ash content (Holmes, 1976), possibly because the higher operating temperatures increase the potential for some inorganic elements to be lost through volatization (Paine, 1971). As with most other techniques for body composition analysis, there are no standardized methods for the use of muffle furnaces. A review of the literature revealed drying temperatures ranging from 450 °C (Kaufman & Kaufman, 1975), 500 °C (Lochmiller *et al.*, 1983; Studier *et al.*, 1994), 550 °C (Worthy & Lavigne, 1983; Fiorotto *et al.*, 1987; Thompson & Drobney, 1996), 600 °C (Odum *et al.*, 1965; Fleharty *et al.*, 1973; Jurgens, 1988) and 620 °C (Ewing *et al.*, 1970; O'Farrell & Studier, 1973) and combustion times from 2 (Schreiber & Johnson, 1975; Lochmiller *et al.*, 1983; Jurgens, 1988), 4 (Paine, 1971; Kaufman & Kaufman, 1975), 6 (Fleharty *et al.*, 1973; Studier *et al.*, 1994), and 8 hours (Ewing *et al.*, 1970). Pierson and Stack (1988) provide one of the few comparisons of ashing temperatures and times, and conclude that 500 °C for 5 hours provided the most accurate measurements of ash content.

Neither muffle furnace ashing or bomb calorimetry provides elemental analysis of the inorganic constituent of body composition. Although the details of elemental analysis are beyond the scope of this chapter, techniques such as atomic absorption spectrophotometry (for calcium, magnesium and iron) and flame spectrophotometry (sodium and potassium) have been used to determine the elemental composition of animals (Studier *et al.*, 1994; Studier & Kunz, 1995).

Ash-free lean dry mass (protein and carbohydrate content)

Ash-free lean dry mass (AFLDM) is composed of both protein and carbohydrates. The protein content of various animals has been shown to vary from 10–15% of total body mass (Widdowson, 1950; Raveling, 1979). Carbohydrates represent a much smaller compartment, at less than 1% of the total body mass (Robbins, 1993). Thus, the AFLDM compartment is 91–94% protein. Despite being the dominant component, it is inappropriate to treat protein content as being synonymous with AFLDM, as was done by Thompson and Drobney (1996). AFLDM can be measured indirectly by subtracting the ash mass determined by combustion from the lean dry mass obtained through fat extraction.

Protein content is most commonly measured using the Kjeldahl procedure, in which a sample is boiled in a concentrated sulfuric acid–catalyst solution until all the organic matter is destroyed and the nitrogen has been converted to ammonium sulfate. The solution is cooled, sodium hydroxide is added, and the ammonia is volatized and captured in a weak acid solution and back-titrated with a known-normality acid to determine the nitrogen content of the original sample (Robbins, 1993). Using a mean nitrogen content of 16% for proteins, the nitrogen estimate obtained in the sample can be multiplied by 6.25 (100/16) to obtain a crude protein content (Robbins, 1993). Other techniques involve microdistillation of the sample using a boric acid trap (Hayward, 1965), and colorimetric techniques and nesslerization (Treybig & Haney, 1983; Studier *et al.*, 1994). The colorimetric technique is faster than the Kjeldahl procedure (Treybig & Haney, 1983), although there are no known studies that compare the relative accuracy and precision of these techniques. Once the protein content is measured, carbohydrate estimates can be obtained by subtracting the protein content from the AFLDM content. Other techniques for measuring carbohydrates are detailed by Dowgiallo (1975). For plant samples, carbohydrates need to be further divided into crude fibre and nitrogen-free extract (Jurgens, 1988).

REFERENCES

Angervall, L. & Carltrom, E. (1963). Theoretical criteria for the use of relative organ weights and similar ratios in biology. *Journal of Theoretical Biology*, **4**, 254–9.

Arnold, T.W. & Ankney, C.D. (1997). The adaptive significance of nutrient reserves to breeding American coots: a reassessment. *The Condor*, **99**, 91–103.

Arnould, J.P.Y., Boyd, I.L. & Speakman, J.R. (1996). Measuring the body composition of Antartic fur seals (*Arctocephalus gazella*): validation of hydrogen isotope dilution. *Physiological Zoology*, **69**, 93–116.

Atchley, W.R., Gaskins, C.T. & Anderson, D. (1976). Statistical properties of ratios. I. empirical results. *Systematic Zoology*, **25**, 137–48.

Atkinson, S.N., Nelson, R.A. & Ramsay, M.A. (1996). Changes in body composition of fasting polar bears (*Ursus maritimus*): the effect of relative fatness on protein conservation. *Physiological Zoology*, **69**, 304–16.

Bachman, G. (1994). Food restriction effects on the body composition of free-living ground squirrels, *Spermophilus beldingi*. *Physiological Zoology*, **67**, 756–70.

Bligh, E.G. & Dyer, W.J. (1959). A rapid method of total lipid extraction and purification. *Canadian Journal of Biochemistry and Physiology*, **37**, 911–17.

Brittas, R. & Marcstrom, V. (1982). Studies in willow grouse, *Lagopus lagopus* of some possible measures of condition in birds. *Ornis Fennica*, **59**, 157–68.

Cohen, J. (1977). *Statistical Power Analysis for the Behavioral Sciences*. New York: Academic Press.

Dobush, G.R., Ankney, C.D. & Krementz, D.G. (1985). The effect of apparatus, extraction time, and solvent type on lipid extractions of snow geese. *Canadian Journal of Zoology*, **63**, 1917–20.

Dowgiallo, A. (1975). Chemical composition of an animal's body and of its food. In *Methods for Ecological Bioenergetics*, ed. W. Grodzinski, R.Z. Klekowski & A. Duncan, pp. 160–84. IBP Handbook No. 24, Oxford, UK: Blackwell Scientific Publications, 367 pp.

Ewing, W.G., Studier, E.H. & O'Farrell, M.J. (1970). Autumn fat deposition and gross body composition in three species of *Myotis*. *Comparative Biochemistry and Physiology*, **36**, 119–29.

Fiorroto, M.L., Cochran, W.J., Funk, R.C., Sheng, H-P. & Klish, W.J. (1987). Total body electrical conductivity measurements: effects of body composition and geometry. *American Journal of Physiology*, **252**, R794–R800.

Fleharty, E.D., Krause, M.E. & Stinnett, D.P. (1973). Body composition, energy content, and lipid cycles of four species of rodents. *Journal of Mammalogy*, **54**, 426–38.

Frisch, R.E. (1987). Body fat, menarche, fitness and fertility. *Human Reproduction*, **2**, 521–33.

Galster, W. & Morrison, P. (1976). Seasonal changes in body composition of the arctic ground squirrel, *Citellus undulatus*. *Canadian Journal of Zoology*, **54**, 74–8.

Gyug, L.W. & Millar, J.S. (1980). Fat levels in a subarctic population of *Peromyscus maniculatus*. *Canadian Journal of Zoology*, **58**, 1341–6.

Hayward, J.S. (1965). The gross body composition of six geographic races of *Peromyscus*. *Canadian Journal of Zoology*, **43**, 297–308.

Holmes, R.T. (1976). Body composition, lipid reserves and caloric densities of summer birds in a northern deciduous forest. *American Midland Naturalist*, **96**, 281–90.

Jameson, E.W. & Mead, R.A. (1964). Seasonal changes in body fat, water and basic weight in *Citellus lateralis*, *Eutamias speciosus*, and *E. amoenus*. *Journal of Mammalogy*, **45**, 359–65.

Jenni-Eiermann, S. (1989). Body composition of starved tufted ducks *Aythya fuligula*, pochards *A. ferina*, and little grebes *Tachybaptus ruficollis*. *Wildfowl*, **40**, 99–105.

Jurgens, M.H. (1988). *Animal Feeding and Nutrition*. Dubuque, IA: Kendall/Hunt Publishing Company, 626 pp.

Kaufman, D.W. & Kaufman, G.A. (1975). Caloric density of the old-field mouse during postnatal growth. *Acta Theriologica*, **20**, 83–95.

Kerr, D.C., Ankney, C.D. & Millar, J.S. (1982). The effect of drying temperature on extraction of petroleum ether soluble fats of small birds and mammals. *Canadian Journal of Zoology*, **60**, 470–2.

Kiell, D.J. & Millar, J.S. (1980). Reproduction and nutrient reserves of arctic ground squirrels. *Canadian Journal of Zoology*, **58**, 416–21.

Kunz, T.H., Wrazen, J.A. & Burnett, C.D. (1998). Changes in body mass and body composition in pre-hibernating little brown bats (*Myotis lucifugus*). *Ecoscience*, **5**, 8–17.

Ledger, H.P. (1968). Body composition as a basis for a comparative study of some East African mammals. *Symposia of the Zoological Society of London*, **21**, 289–310.

Licht, P. & Leitner, P. (1967). Physiological responses to high environmental temperatures in three species of microchiropteran bats. *Comparative Biochemistry and Physiology*, **22**, 371–87.

Lochmiller, R.L., Whelan, J.B. & Kirkpatrick, R.L. (1983). Body composition and reserves of energy of *Microtus pinetorum* from Southwest Virginia. *American Midland Naturalist*, **110**, 138–44.

Lynch, G.R. (1973). Seasonal changes in thermogenesis, organ weights, and body composition in the white-footed mouse, *Peromyscus leucopus*. *Oecologica*, **13**, 363–79.

Mascher, J.W. & Marcstrom, V. (1976). Measures, weights, and lipid levels in migrating dunlins *Calidris alpina* at the Ottenby Bird Observatory, South Sweden. *Ornis Scandinavica*, **7**, 49–59.

McNab, B.K. (1976). Seasonal fat reserves of bats in two tropical environments. *Ecology*, **57**, 332–8.

Meijar, Möhring, F.J. & Trillmich, F. (1994). Annual and daily variation in body mass and fat of Starlings *Sturnus vulgaris*. *Journal of Avian Biology*, **25**, 98–104.

Millar, J.S. (1975). Tactics of energy partitioning in breeding *Peromyscus*. *Canadian Journal of Zoology*, **53**, 967–76.

Millar, J.S. (1981). Body composition and energy reserves of northern *Peromyscus leucopus*. *Journal of Mammalogy*, **62**, 786–94.

Miller, I. & Weil, W.B. (1963). Some problems in expressing and comparing body composition determined by direct analysis. *Annals of the New York Academy of Science*, **110**, 153–60.

Morton, M.L. (1975). Seasonal cycles of body weights and lipids in belding ground squirrels. *Bulletin of the Southern California Academy of Sciences*, **74**, 128–43.

Myrcha, A. & Walkowa, W. (1968). Changes in the caloric value of the body during the postnatal development of white mice. *Acta Theriologica*, **13**, 391–400.

Nestlet, J.R, Dieter, G.P. & Klokeid, B.G. (1996). Changes in total body fat during daily torpor in deer mice (*Peromyscus maniculatus*). *Journal of Mammalogy*, **77**, 147–54.

Odum, E.P., Marshall, S.G. & Marples, T.G. (1965). The caloric content of migrating birds. *Ecology*, **46**, 901–4.

O'Farrell, M.J. & Studier, E.H. (1973). Reproduction, growth, and development in *Myotis thyanodes* and *M. lucifugus* (Chiroptera: Vespertilionidae). *Ecology*, **54**, 18–30.

Paine, R.T. (1971). The measurement and application of the calorie to ecological problems. *Annual Review of Ecology and Systematics*, **2**, 145–64.

Piersma, T. (1988). Body size, nutrient reserves and diet of Red-necked and Slavonian Grebes *Podiceps grisegena* and *P. auritus* on Lake Ijsselmeer, The Netherlands. *Bird Study*, **35**, 13–24.

Pierson, E.D. & Stack, M.H. (1988). Methods of body composition analysis. In *Ecological and Behavioral Methods in the Study of Bats*, ed. T.H. Kunz, pp. 387–403. Washington, DC: Smithsonian Institute Press. 533 pp.

Raveling, D.G. (1979). The annual cycle of body composition of Canada geese with special reference to control of reproduction. *The Auk*, **96**, 234–52.

Reinecke, K.J. & Stone, T.L. (1982). Seasonal carcass composition and energy balance of female black ducks in Maine. *The Condor*, **84**, 420–6.

Reynolds, D.S. & Kunz, T.H. (2000). Changes in body composition during reproduction and postnatal growth in the little brown bat, *Myotis lucifugus* (Chiroptera: Vespertilionidae). *Ecoscience*, **7**, 10–17.

Rickart, E.A. (1982). Annual cycles of activity and body composition in *Spermophilus townsendii mollis*. *Canadian Journal of Zoology*, **60**, 3298–306.

Ringelman J.K. & Szymczak, M.R. (1985). A physiological condition index for wintering mallards. *Journal of Wildlife Management*, **49**, 564–8.

Robbins, C.T. (1993). *Wildlife Feeding and Nutrition*. New York: Academic Press. 343 pp.

Sawicka-Kapusta, K. (1975). Fat extraction in the Soxhlet apparatus. In *Methods for Ecological Bioenergetics*, ed. W. Grodzinski, R.Z. Klekowski & A. Duncan, IBP Handbook No. 24, pp. 228–92. Oxford: Blackwell Scientific Publications. 367 pp.

Schreiber, R.K. & Johnson, D.R. (1975). Seasonal changes in body composition and caloric content of Great Basin rodents. *Acta Theriologica*, **20**, 343–64.

Schmidt-Nielsen, K. (1994). *Animal Physiology*. New York: Cambridge University Press. 602 pp.

Sokal, R.R. & Rohlf, F.J. (1982). *Biometry*. New York: W.H. Freeman and Company. 859 pp.

Studier, E.H. & Kunz, T.H. (1995). Accretion of nitrogen and minerals in suckling bats, *Myotis velifer* and *Tadarida brasiliensis*. *Journal of Mammalogy*, **76**, 32–42.

Studier, E.H., Sevick, S.H. & Wilson, D.E. (1994). Proximate, caloric, nitrogen and mineral composition of bodies of some tropical bats. *Comparative Biochemistry and Physiology*, **109A**, 601–10.

Thompson, J.E. & Drobney, R.D. (1996). Nutritional implications of molt in male canvasbacks: variation in nutrient reserves and digestive tract morphology. *The Condor*, **98**, 512–26.

Tidemann, C.R. (1982). Sex differences in seasonal changes of brown adipose tissue and activity of the Australian vespertilionid bat *Eptesicus vulturnus*. *Australian Journal of Zoology*, **30**, 15–22.

Treybig, D.S. & Haney, P.L. (1983). Colorimetric determination of total nitrogen in amines with selenium catalyst. *Analytical Chemistry*, **55**, 983–5.

Webb, P.I. (1992). Aspects of the ecophysiology of some vespertilionid bats at the northern borders of their distribution. PhD thesis, University of Aberdeen, Aberdeen, Scotland, 150 pp.

Weber, M.L. & Thompson, J.M. (1998). Seasonal patterns in food intake, live mass, and body composition of mature female fallow deer (*Dama dama*). *Canadian Journal of Zoology*, 76, 1141–52.

Widdowson, E.M. (1950). Chemical composition of newly born mammals. *Nature*, **166**, 626–8.

Winstanley, R.K., Saunders, G. & Buttemer, W.A. (1998). Indices for predicting total body fat in red foxes from Australia. *Journal of Wildlife Management*, **62**, 1307–12.

Winstanley, R.K., Buttemer, W.A. & Saunders, G. (1999). Fat deposition and seasonal variation in body composition of red foxes (*Vulpes vulpes*) in Australia. *Canadian Journal of Zoology*, **77**, 406–12.

Wishart, R.A. (1979). Indices of structural size and condition of American wigeon (*Anas americana*). *Canadian Journal of Zoology*, **57**, 2369–74.

Worthy, G.A.J. & Lavigne, D.M. (1983). Changes in energy stores during postnatal development of the harp seal. *Journal of Mammalogy*, **64**, 89–96.

JOHN R. SPEAKMAN, GEORGE H. VISSER, SALLY WARD
AND ELŻBIETA KRÓL

3

The isotope dilution method for the evaluation of body composition

Introduction

One key to measuring the body composition of an animal without killing it in the process is the fact that water is not evenly distributed in body tissues. Fat contains substantially less water than lean tissue and this difference means that the fatter an organism becomes, the lower the water content as a percentage of its total body mass. Since body mass is relatively easily measured, if the total water content of an animal could also be quantified, a method would be available for estimating fatness. Several of the methods detailed in this book (for example, TOBEC: Chapter 5 and BIA: Chapter 6) rely on the differential water contents of lean and fat tissue to quantify body composition. Dilution methods are also based on this principle. Initial attempts to measure the body water by dilution used compounds that were soluble in water (such as antipyrene: e.g. Soberman *et al.*, 1949, urea: e.g. Meissner, 1976 or thiocyanate: e.g. Hollander *et al.*, 1949). The discovery of isotopes of oxygen and hydrogen in the 1920s and 1930s had opened up the opportunity of using these materials to 'label' the body water directly, and the first attempts to do this were made in the 1930s (von Hevesy & Hofer, 1934). The isotope dilution method grew out of these initial studies.

Theory

To understand how the isotope dilution method works, it is perhaps useful to consider an analogous situation with which many animal ecologists will be familiar: the problem of determining the size of a population of animals that live in a given area. Estimation of total body water may not, at first sight, appear similar to the problem of counting animals. However, if it is considered that the question is one of estimating the number of water molecules within a body, the problems are actually very similar. In both cases we are dealing with a defined number of objects (animals or molecules) in a defined space (an area of land or a body). Moreover, in both circumstances the objects are not completely restricted to remain within the defined space, and over time they will drift into and out of the space we have delineated, although at any particular time the number within the space is relatively stable.

One way of counting the number of animals in an area might be to go through the area and shoot them all. This would certainly be effective and yield an accurate and precise answer but the effect would be to destroy the thing one is attempting to measure. The analogous process for body composition analysis is to kill the animal and perform a chemical analysis (see Chapter 2) and its effects are similarly accurate, precise and disastrous for the animal in question.

A useful technique employed by many population biologists to non-invasively estimate the size of a population is the mark/recapture technique (Seber, 1982). This technique involves first capturing a subset of the entire population. These animals are marked with a tag which cannot be removed, such as a band around a limb, or more recently a popular method has been to use a subcutaneously implanted transponder. The marked animals are released back into the population and then on a later occasion a sample of the entire population is recaptured. In this recaptured sample only some of the original marked animals will be present (unless the entire population was captured first time around). From the original number of animals that were marked and the proportion of marked animals in the recaptured sample, an estimate can be made of the population size (Seber, 1982). The larger the population relative to the number of animals that are marked, the more the marked individuals will be 'diluted' and their representation in the second sample will diminish. It is rather simple to show that given a population of N individuals and a number of marked individuals M that:

$$N = M/pM \tag{1}$$

where pM is the proportion of marked individuals in the second sample.

The isotope dilution method works in almost exactly the same way. There are N molecules of water in the body. A certain number of 'marked' water molecules are introduced into the body. The marking is generally done by replacing one of the hydrogen molecules with a heavy isotope of hydrogen – deuterium ^2H or tritium ^3H or the oxygen molecule with a heavy isotope of oxygen – ^{17}O or ^{18}O. Later, a sample of body water is taken to examine the extent to which the marked water molecules have been diluted in the total population of water molecules – by measuring the isotopic enrichment in the sample. As we cannot count the number of 'marked' water molecules introduced into the body directly, we can only estimate the number from their mass, and the known molecular mass of the marked molecules. Hence, instead of the number of marked molecules (M) we introduce a mass of marked molecules M_m which is equivalent to a given number of mols of these molecules M_{mol}

$$M_{mol} = M_m / MM_m \tag{2}$$

where MM_m is the molecular mass of the marked molecules. By analogy with eqn. 1

$$N_{mol} = (M_{mol}) / pM \tag{3}$$

To estimate the 'population' of water molecules in the body therefore, one needs to introduce a known mass of marked water molecules and then at some later time take a sample of body water and examine the proportion of molecules which are marked (pM). Generally, this measurement is made as an enrichment E of the given isotope in ppm and expressed as a proportion by dividing by 10^6. The resultant proportional contribution of the isotope is often called the atom percent (AP), and when this is expressed as the amount above background (see below) it is termed atom percent excess (APE).

This simple analogy with mark/recapture methods allows us to explore some of the assumptions which attend the use of the isotope dilution method, because the assumptions upon which the mark/recapture method is founded have direct analogies with the assumptions of the isotope dilution method. Understanding problems with measuring populations of animals, however, perhaps provides a convenient and familiar route for field biologists to appreciate some of the subtleties of determining body composition by isotope dilution.

Assumptions of the mark/recapture and isotope dilution methods

The first assumption when using the mark/recapture method is that the marked animals disperse themselves evenly through the entire population.

It would be no use, for example, if the marked animals all congregated in a subpopulation of the whole population in which we are interested. If this did happen and we were to go out and recapture at a later date, we would either massively overestimate the population if by chance we failed to sample this subpopulation, or we would massively underestimate the population if we recaptured predominantly from the subpopulation. The method by which the marked individuals are introduced back into the population might therefore be critical, as well as the time left for the animals to adequately disperse.

The population of water molecules within the body is composed of sub-populations that do not mix completely and instantaneously. Hence, if a population of marked water molecules is introduced into the blood, it will be diluted only in the blood at first and then the marked molecules will more slowly pervade the rest of the system, which consists primarily of water between cells (interstitial water) and water within cells. These different populations of water molecules are commonly called the body water pools. It is necessary, when using the isotope method, to allow sufficient time for the marked molecules to pervade the entire body water. This, in turn, will depend on the route by which the marked molecules are introduced. If the isotope is introduced intravenously (i.v.), it will pervade the blood very rapidly and then slowly spread into the surrounding pools. A sample of blood taken shortly after injection would contain a very high isotope enrichment because most of the isotope would still be circulating in the blood system, e.g. see figures in Edelman (1952). This would lead to an underestimate of the total body water pool size because one would effectively be determining only the blood volume. In contrast, if the dose was administered orally, the uptake of the isotope from the gut would be relatively slow, and the transfer of isotope from the gut into the blood would not greatly exceed the rate at which the isotope perfused the other body pools. In consequence, the pattern of isotope enrichment in blood post administration would slowly converge on the plateau enrichment from below, e.g. Wong *et al.* (1988).

These mixing processes can be relatively easily modelled by considering that the body consists of two pools: one pool is smaller than the other and it is this pool which takes up the isotope when it is administered (Fig. 3.1(*a*)). The isotope initially remains in its own pool (V_d) and is subsequently incorporated into the small body pool (V_s). The system is characterized by the sizes of the three reservoirs (V_d, V_s and V_L) and the rates at which material moves between them (characterized by exponential

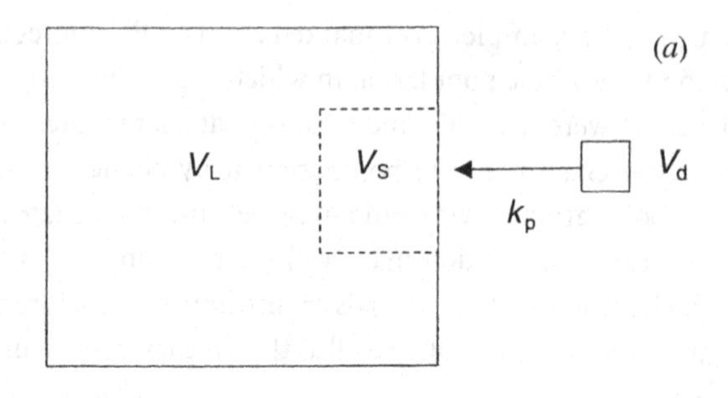

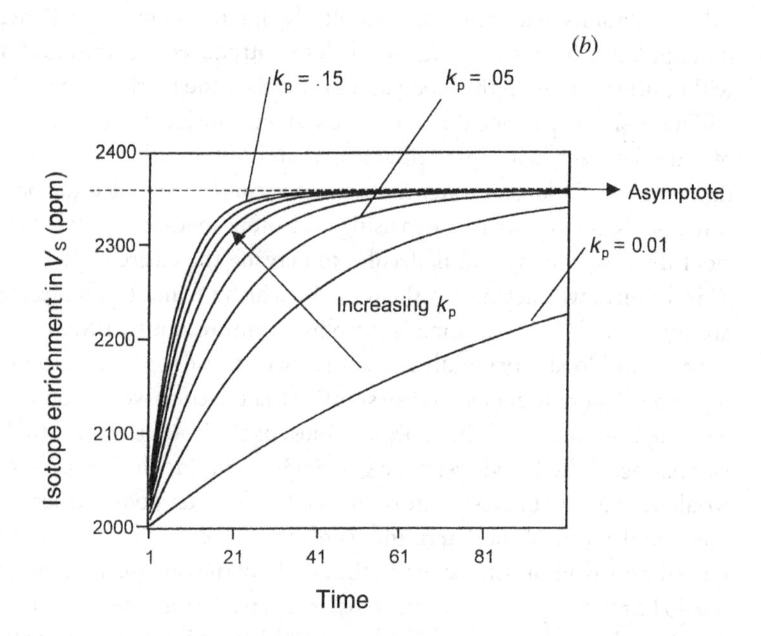

Fig. 3.1. Patterns of isotope enrichment (following dosage) of a body consisting of two distinct pools (V_S and V_L) of isotope (V_d). In Fig. 3.1(*a*) the large and small pools mix instantaneously and therefore act functionally as a single pool. The rate at which the dose penetrates the pool is described by an exponential parameter (k_p). Fig. 3.1(*b*) shows the patterns of isotope enrichment as a function of time (arbitrary units) following dosing for different values of k_p. As k_p increases, the speed at which the dose equilibriates with the body increases.

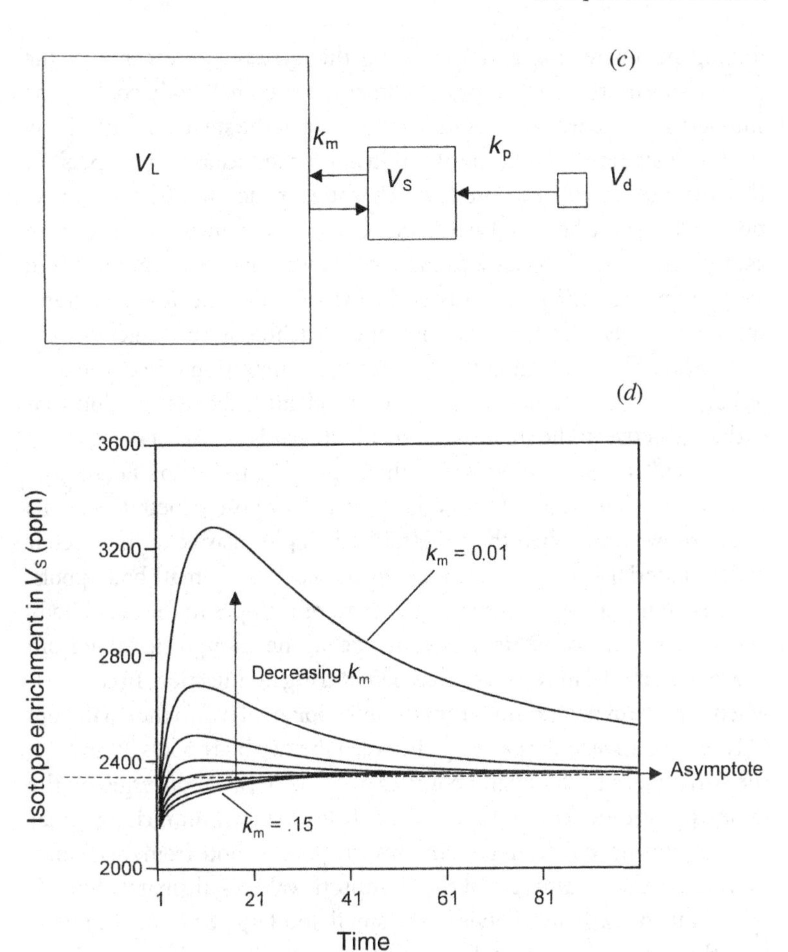

Fig. 3.1(*c*) illustrates a model where V_S and V_L do not mix instantaneously but interact at a rate defined by a mixing parameter k_m. Fig. 3.1(*d*) shows the patterns of variation in isotope enrichment in the small pool (V_S) as a function of time following dosing. In this example k_p is set at 0.1 and k_m is allowed to vary between 0.01 and 0.15. The curves again converge on an asymptotic value, but the ratio of k_m to k_p defines whether this convergence occurs from above or below.

mixing parameters: k_p and k_m). Using this system, we can model the pattern of variation in isotope enrichment in the small body pool. In the simplest scenario we can imagine that mixing of the small and large body pools is instantaneous (k_m high) and in effect they act as a single pool. In this instance, varying the rate at which isotope penetrates from the injection pool into the body (k_p) yields a series of curves which converge on an asymptotic value from below as the dose is successively incorporated into the system (Fig. 3.1(*b*)). The greater the rate at which the dose penetrates the body pools, the faster the isotope enrichment converges on the asymptotic value. By fixing the value of the penetration rate of the dose (k_p) at 0.1, we can explore the impact of relaxing the assumption that exchange between the small and larger body pools is instantaneous (Fig. 3.1(*c*) and (*d*). In this case we varied the value of k_m from 0.01, i.e. 1/10 the rate of dose penetration to 0.15, i.e. faster than dose penetration. This figure shows that, when the penetration is rapid relative to the rates at which material is exchanged between the large and small body pools, there is an initial overshoot in enrichment of isotopes in the small body pool (V_s). The enrichments then converge on the asymptotic value from above. This model mimics what happens during i.v. injection. In contrast, when the rate of penetration from the injection pool (V_d) to the small pool (V_s) is low relative to the exchange between the two body pools (V_s and V_L), the enrichment in the small pool rises only slowly, and converges on the same asymptotic equilibrium value from below, mimicking what happens during oral dosing. An interesting point to note from this figure, however, is that attainment of the asymptotic value is ultimately limited by the rate of exchange between the small and large body pools, rather than the penetration rate of the dose. Fast penetration by the dose will not inevitably lead to a shorter equilibrium time in the whole body water.

The second assumption when using mark/recapture is that the timing of the recapture must be made after a sufficient period to enable the marked individuals to disperse through the population, but it must also not be so long that marked individuals have had a chance to leave (or die) and be replaced by unmarked immigrants and newborn animals. With the mark/recapture method the timescale over which dispersal occurs within a population is generally very rapid relative to the demographic processes of immigration/emigration and fecundity/mortality. However, the total body water pool is in a state of constant flux because there are continual outflows in the form of urine, water in faeces, and evaporative processes (equivalent to death and emigration) and inevitably these will remove

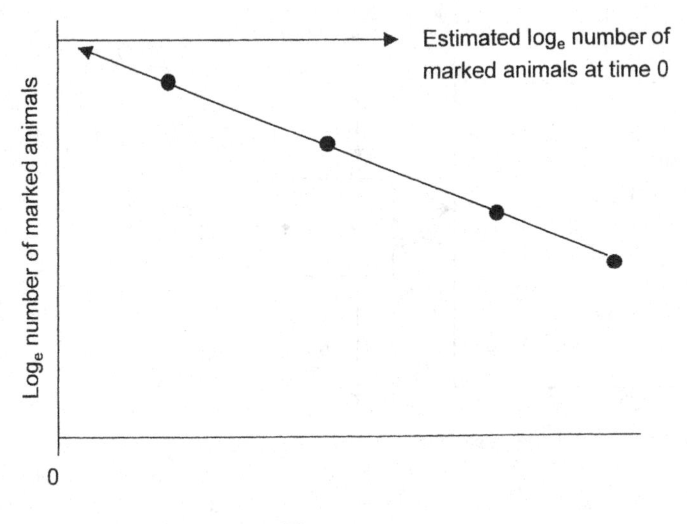

Fig. 3.2. Hypothetical pattern of change in the population of a marked group of animals over time due to the processes of death and emigration. By back-extrapolating this curve to the time when the animals were initially marked (time 0), it is possible to assess the initial numbers of marked animals as a percentage of the total population.

some of the marked molecules, which will be replaced by the continual inputs of water (from drinking, preformed water in food and from the continuous formation of metabolic water). These latter inputs would be equivalent to immigration and natality of unmarked individuals. These processes may not occur at rates which can be ignored relative to the time necessary for the marked molecules to disperse through the body.

In the mark/recapture protocol, the way in which losses of marked individuals due to emigration and mortality can be accounted for is to make not just a single recapture but to make several recaptures over a period of time. If mortality and emigration are occurring, the number of marked individuals will gradually decline over time, as marked individuals are gradually replaced by unmarked individuals. One could then reconstruct the proportion of marked individuals at the time they were released by fitting a curve to this decline and back-extrapolating to the point at which the marked individuals were introduced to the population (Fig. 3.2).

We can model what is happening in this situation by including a flow of isotope into the small pool (V_S) from the environment and a sink away from this pool (Fig. 3.3(*a*)). Using this more refined system, we explored the

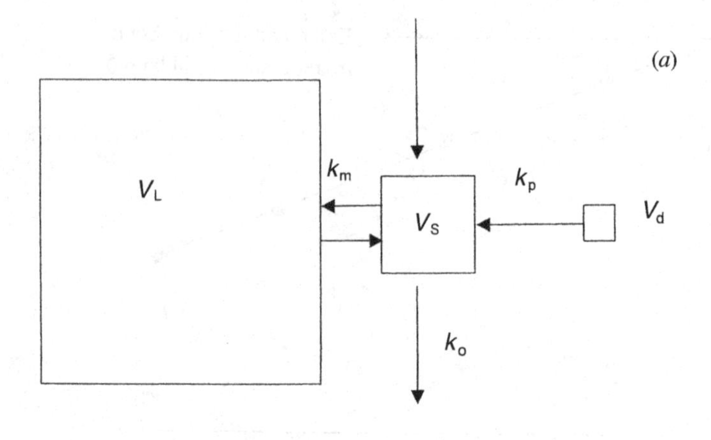

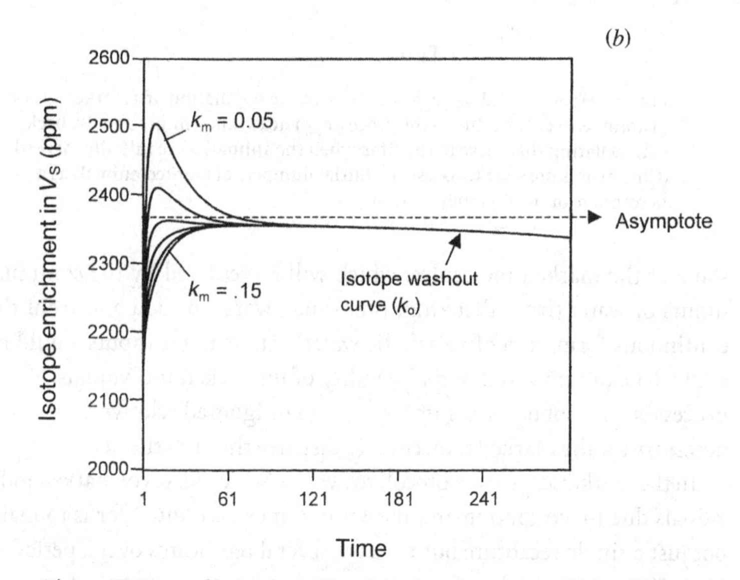

Fig. 3.3. Patterns of isotope equilibration when there is an additional flow of material through V_S (described by k_o) illustrated by the model in Fig. 3.3(*a*). In Fig. 3.3(*b*) the value of k_o is held constant at 0.01 and the value of k_m is allowed to vary between 0.05 and 0.15. In this case, the tracks are similar to those in Fig. 3.1(*d*) but, instead of converging on the asymptote, they converge on the elimination line. Beyond 90 time units, the curve is completely dominated by the value of k_o.

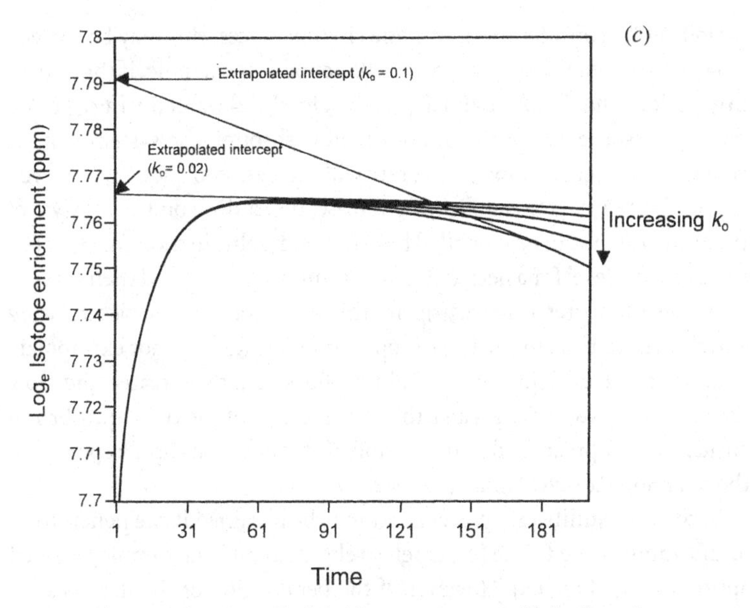

In Fig. 3.3(*c*) the value of k_m is held constant at 0.1 and k_o is allowed to vary between 0.01 and 0.02. If these different elimination curves are back-extrapolated to time 0, they do not predict the same initial start point. Greater turnovers (high k_o) lead to higher values of the back-extrapolated intercept.

patterns of isotope enrichment post-injection for variations in the rate of exchange between the large and small pools (k_m) and the flow through the system (k_o) (Fig. 3.3(*b*)). Now the isotope enrichments do not converge on the asymptotic value but rather home in on an overall decline in the isotope enrichment over time which is superimposed upon the original pattern. The labelling isotope never reaches a steady state equilibrium value because it is being constantly lost. In theory, however, the curves relating enrichment over time following the isotope administration extrapolate back to the equilibrium value at the time of administration, i.e. the asymptote (Fig. 3.3(*b*)). Using this method of back-extrapolation is generally called the 'intercept' approach. An important point to note from Fig. 3.3(*b*), however, is that the back-extrapolation process only works if the initial isotope samples in the series are taken after the mixing processes have been completed. Taking earlier samples would result in erroneous estimates of the elimination gradient and hence erroneous back-extrapolations.

This theoretical use of the intercept makes the assumption that the flow through process occurs at the same rate throughout the entire period under consideration. In some circumstances, this may be correct. However, in some circumstances it will not be. For example, if the animal in question is held captive and deprived of food and water during a period following isotope administration, the flow through the system may be very low (Calazel *et al.*, 1993). The rate will increase enormously, however, and to different extents, following release, depending on the lifestyle of the animal. This can be modelled by keeping the elimination constant (k_o) fixed at a low level for a period during the mixing phase, and then varying its value thereafter (Fig. 3.3(*c*)). In this situation, the curves relating enrichment to time following isotope administration do not extrapolate back to the true equilibrium value. Moreover, the greater the flow through the system the greater the overestimate of the true equilibrium enrichment. A practical demonstration of this effect has been reported in the red knot (*Calidris canutus*) (Visser *et al.*, 2000).

A correct equilibrium enrichment may be obtained if the penetration of the administered dose (k_p) is relatively rapid and the sample is timed appropriately (Fig. 3.3). However, if the penetration of the dose is slow (k_p low) and mixing of the body pools relatively rapid (k_m high), the equilibrium enrichment may never be adequately estimated – because isotope is being washed out of the system at a significant rate relative to the rate at which it is trying to flood into it. Relying on a single sample where the processes of dispersal of the marker and losses from the system as a whole are presumed to be in equilibrium is termed the 'plateau' approach.

The third assumption when using the mark/recapture method is that individuals are unable to shed their marks and that the population consists initially of completely unmarked individuals. Usually, tag loss is the more important of these problems for mark/recapature since the likelihood of finding individuals that are already marked is remote. In contrast, the utilization of stable isotopic markers on water molecules means that transmutation of the marks on molecules can be ignored (although using radioactive markers such as tritium is effectively analogous to tag loss because radioactive markers can spontaneously decay and become 'unmarked'). This might possibly appear to favour the use of stable over radioactive markers. There are, however, two problems with this interpretation. First, the rates of radioactive decay are infinitessimally small relative to the durations over which isotope dilution measurements are made. The half-life of tritium for example is 12.3 years, while the duration of a typical dilution experiment would normally be a few hours. Tritium, however, occurs naturally at levels that can be effectively ignored. In contrast, stable isotope markers (deuterium, ^{17}O and ^{18}O) are found naturally within the body and consequently there is a background level of marked molecules which must be accounted for when utilizing the isotope method. In general, if the enrichment of the isotope in the material introduced into the animal is termed E_{in}, the background level of this enrichment in the animal is E_b, and the enrichment measured after the 'dispersal' process is completed is E_p (*p* for plateau enrichment), the number of moles of water present in the body (N_{mol}) can be estimated as

$$N_{mol} = \frac{M_{mol}(E_{in} - E_p)}{(E_p - E_b)} \quad (4)$$

Eqn. 4 is derived by rearranging eqn. 17.1 in Speakman (1997). The enrichment estimated at the intercept can be substituted for the plateau enrichment (E_p) if the intercept approach is used rather than the single sample method. The presence of background levels of the markers means it is sometimes necessary to establish the level of E_b prior to introducing the marked population (see Practical aspects below for details).

The fourth assumption is that marked and unmarked individuals behave in identical manners within the population – marking them does not cause them to behave in different ways or for example be exposed to differential amounts of predation and emigration. Early attempts to measure the amount of water within the body using the dilution principle relied on using compounds that dissolve readily in water, such as antipyrene (Kay *et al.*, 1966; Panaretto, 1963a; Panaretto & Till, 1963; Reid *et al.*, 1958; Soberman *et al.*, 1949), urea (Andrew *et al.*, 1995; Meissner, 1976) or thiocyanate (Freis *et al.*, 1953; Hollander *et al.*, 1949). One potential problem with these markers is that they differ chemically from water and may not therefore behave in the same manner, for example, by being eliminated in different ways or not dispersing in the body completely into all the inter- and intracellular pools. Replacing these dissolved compounds with isotopes effectively removed these problems because one was actually marking the water molecules themselves. However, the solution is not perfect because, while isotopically labelled water molecules behave similarly to unlabelled molecules, they do not do so exactly. The differences in behaviour of the molecules which are labelled and unlabelled stem mostly from the fact that the molecules have slightly different molecular masses, and molecular mass influences the physical characteristics of the molecule. Heavier molecules generally require greater amounts of energy to take part in reactions or physical processes such as evaporation. For example, the rates at which tritium is incoprorated into fat molecules during *de novo* fat synthesis is between 15 and 20% lower than the rate at which deuterium is incorporated (Jungas, 1968; Eidinof *et al.*, 1953; Glascock & Duncombe, 1952). Similar effects are also apparent in the different rates at which evaporation occurs during loss of water from the body, because the heavier molecules containing isotopic labels require greater activation energies to evaporate and consequently they are underrepresented in the departing population by between 1 and 10% (Kirshenbaum, 1951).

The most direct consequences of these differences in isotope behaviour is that different labelling of molecules might be expected to take different times to mix throughout the body. Our knowledge of the diffusion processes of the isotopes, however, indicates that differences are likely to be relatively minor (von Hevesy & Jacobson, 1940; Chinard & Enns, 1954; Enns & Chinard, 1956) and this effect can probably be ignored in practical situations (but see estimates of differential diffusion rates in Lucke & Harvey, 1935).

The fifth assumption when using the mark/recapture method is that one can easily distinguish members of the population from mimics that are not 'true' population members. In 1991 we summarized a body of work which had aimed over the previous two decades to estimate the population size of several bat species which live in the Aberdeen area (Speakman *et al.*, 1991). It later transpired that one of these species was two sibling species which morphologically are almost identical but were demonstrated to have different echolocation calls in 1993 (Jones & van Parijs, 1993) and were shown to be genetically distinct in 1997 (Barratt *et al.*, 1997). In our area, one of the sibling species is quite rare, but the population size for the other much commoner species was inflated by the presence of these individuals.

In the body the components of the water molecules are assumed to remain parts of water molecules. By marking some of the hydrogen or oxygen atoms on these molecules, the observed dilution is assumed to reflect the number of water molecules in the body. However, the oxygen and hydrogen in water molecules are not static and they exchange with other hydrogen and oxygen molecules in the body. Hydrogen and oxygen are not therefore true markers for water molecules; what they actually measure are the populations of exchangeable hydrogen and oxygen in the body. In effect, there are small extra populations of molecules that the dilution method includes in the 'total body water' estimate. We assume the dilution effect is due to water, but this leads to a slight error in the same way that we could not separate the common and rare forms of the bats living around Aberdeen and thus overestimated the population of the commoner form. These extra populations are different for oxygen and hydrogen because hydrogen exchanges much more readily with other hydrogens in the body. On average the extra hydrogen space is 4.59% larger than the body water pool estimated by dessication in mammals and 4.73% in birds (Table 3.1). These extra spaces are not connected to body mass in any obvious manner but the interspecific variations in the extent to which the hydrogen space differs from the total body water pool size are large, ranging from underestimates of 2.1% in the mule deer (Torbit *et al.*, 1985) to overestimates of 22.5% in the cow (Little & McLean, 1981). Moreover, even within given species the variability can be enormous. The cow provides a convenient example where separate studies have given excess hydrogen pool sizes between 1.5 and 22.5% (averaging 11.5% Table 3.1).

Relatively few studies have made comparisons between the ^{18}O dilution space and the body water pool size by desiccation (Table 3.2). Yet these studies suggest that the ^{18}O space is considerably closer to the body water pool exceeding it by, on average, only 1.017%. Moreover, the different studies appear far more consistent in the extent to which the oxygen space exceeds the body water pool (s.d. = 0.0125, across six species) compared with the variation in the extent to which the hydrogen space exceeds the body water pool (s.d. = 0.053, across 27 species means). The variance (s.d.^{2}) for hydrogen is about 18× greater than for oxygen. This means that isotopic markers of oxygen (^{17}O and ^{18}O) provide more reliable estimates of body water than do isotopic markers of hydrogen (deuterium or tritium). However, water enriched with ^{17}O or ^{18}O costs between 10 and 20× more than deuterated and 30–40× more than tritiated water. The cost of the dose of isotope required to label the animal must therefore be traded off against the accuracy of the individual determination. In a study in which the most accurate data possible is required from individual animals, ^{18}O or ^{17}O should be used, especially if other constraints mean that only a small number of individuals can be labelled. In a study in which the amount of money available to purchase labelling isotopes limits the sample of animals before the number of animals which are available for study, or where groups of animals are to be compared, it may be better to label larger numbers of animals with

Table 3.1. *Hydrogen dilution space expressed relative to the body water volume evaluated by desiccation*

Species	Ratio	*n*	D/T	Reference
Mammals				
Mice				
Mus	1.037		T	Siri and Evers (1962)
	1.064	16	D	Lifson *et al.* (1955)
	1.0437	8	D	McClintock and Lifson (1957a)
(ob/ob)	1.0375	8	D	McClintock and Lifson (1957b)
	1.0455		**Mean**	
Calomys	1.031	5	T	Holleman and Dietrich (1973)
	1.052	5	T	Holleman and Dietrich (1975)
	1.0415		**Mean**	
Acomys	1.023	5	T	Holleman and Dietrich (1975)
Peromyscus	1.023	8	T	Holleman and Dietrich (1975)
Antechinus	1.062		T	Nagy *et al.* (1978)
Rats				
Rattus	1.0657	10	T	Foy and Schneiden (1960)
	1.0647	7	T	
	1.1205	32	T	Tsivipat *et al.* (1974)
	1.11		T	Bell and Stern (1977)
	1.0171	21	T	Culebras *et al.* (1977)
	1.097	40	T	Gordon *et al.* (1971)
	1.064		T	Siri and Evers (1962)
	1.02	4	D	McClintock and Lifson (1958)
	1.0699		**Mean**	
Dipodomys	**1.026**		T	Nagy and Costa (1980)
Vole	**1.001**	4	T	Holleman and Dietrich (1975)
Lemming	**0.943**	5	T	Holleman and Dietrich (1975)
Hamster	**1.0274**	34	T	Kodama (1971)
Pocket Gopher	**1.0916**	12	T	Gettinger (1983)
Antelope-ground squirrel	**1.05**		T	Karasov (1983)
Guinea pig	**1.016**		T	Siri and Evers (1962)
Rabbit	1.0096	9	D	Moore (1946)
	0.9964	2	T	Pace *et al.* (1947)
	1.0049	20	T	Reid *et al.* (1958)
	1.0036		**Mean**	
Bat	**1.125**		T	Inwards, in Ellis *et al.* (1991)
Dog	1.2312	25	T?	Sheng and Huggins (1971)
	1.2090	18	T?	Sheng and Huggins (1971)
	1.2201		**Mean**	
Pig	0.9835	26	D	Groves and Wood (1965)
	1.1151	40	T	Kay *et al.* (1966)
	0.9944	15	D	Flynn *et al.* (1968)
	1.023	24	D	Houseman *et al.* (1973)
	1.0901	55	T	Setiabudi *et al.* (1975)
	1.2021	73	D	Shields *et al.* (1983)
	1.1862	63	D	Shields *et al.* (1984)

Table 3.1 (*cont.*)

Species	Ratio	*n*	D/T	Reference
Pig (*cont.*)				
	1.1766	54	D	Ferrell and Cornelius (1984)
	1.0964		**Mean**	
Cow	1.1683	13	T	Carnegie and Tulloch (1968)
	1.1338	13	T	
	1.096	12	D	Crabtree *et al.* (1974)
	1.0499	20	T	Meissner *et al.* (1980)
	1.225	31	T	Little and McLean (1981)
	1.1497	12	D	Odwongo *et al.* (1984)
	1.0154	28	D	Arnold *et al.* (1985)
	1.0555	21	D	Andrew *et al.* (1995)
	1.1117			**Mean**
Goat	1.0380	13	T	Panaretto and Till (1963)
	1.0405	10	T	Panaretto (1963b)
	1.1087	12	T	Viljoen *et al.* (1988)
	1.0624			**Mean**
Sheep	1.0883	9	T	Panaretto (1963b)
	1.0529	15	T	Panaretto (1968)
	1.088	61	T	Searle (1970)
	1.180	13	D	Foot and Greenhalgh (1970)
	1.163	24	T	Farrell and Reardon (1972)
	1.030		T	Smith and Sykes (1974)
	1.0189	9	T	Meissner (1976)
	1.1290	25	T	Foot *et al.* (1979)
	1.0937			**Mean**
Mule deer	**0.9785**	4	T	Torbit *et al.* (1985)
White-tailed deer	**1.0361**	9	D	Rumpler *et al.* (1987)
Antarctic-fur seal	**1.0192**	9	T	Arnould *et al.* (1996)
	1.0174	5	D	
Ringed seal	**1.016**	2	T	Lydersen *et al.* (1992)
Harp seal	**1.017**	9	D	Oftedal *et al.* (1996)
Grey seal	**1.028**	4	T	Reilly and Fedak (1990)
	1.040	4	D	
Mammals	**1.0459**			
	26 species			
	s.d. = 0.0525 s.e. = 0.01			
Birds				
Pigeon	**1.025**	2	T	Siri and Evers (1962)
Chukar	0.998	8	T	Degen *et al.* (1981)
	1.046	11	T	Crum *et al.* (1985)
	1.022		**Mean**	
Sparrow	**1.044**	9	T	K. A. Nagy (unpublished)[a]
Chicken	1.180	240	T	Farrell (1974)
	1.150	16	T	Farrell and Belnave (1977)
	0.973	19	D	Kirchgessner *et al.* (1977)
	1.104	169	T	Johnson and Farrell (1988)
	1.085	115	D	
	1.0984		**Mean**	

Table 3.1 (*cont.*)

Species	Ratio	n	D/T	Reference
Birds	1.0473			
	4 species			
	s.d. = 0.0354 s.e. = 0.0177			
Reptiles				
Anolis	**1.040**		T	Nagy and Costa (1980)
Chuckwalla	**1.039**		T	Nagy and Costa (1980)
Arthropods				
Tenebrio	1.131		T	Bohm and Hadley (1977)
Scorpion	1.060		T	King (1976)
Bombus	1.004		T	Wolf *et al.* (1996)

Notes:
n refers to the number of individual animals involved in the validation study and D/T refers to the hydrogen isotope used (T = Tritium and D = Deuterium).
[a] cited in Crum *et al.* (1985).

Table 3.2. ^{18}*Oxygen space relative to the body water space evaluated by desiccation*

Species	Ratio	n	Reference
Mammals			
Mus	**1.016**	6	Lifson *et al.* (1955)
Dog	**1.0067**	8	Speakman *et al.* (2000)[a]
Birds			
Sparrow	**1.023**	6	Williams (1985)
Sparrow	**1.0094**	4	Williams and Nagy (1984)
Hummingbird	**1.037**		Tiebout and Nagy (1991)
Sparrow	**1.005**	9	K.A. Nagy (unpublished)[b]
Birds and mammals			
	1.018		
	5 species		
	s.d. = 0.0125 s.e. = 0.006		

Notes:
[a] relative to DXA not desiccation (not included in average).
[b] cited in Crum *et al.* (1985).

one of the hydrogen isotopes. The cost of labelled isotope sample analysis and availability of equipment required to analyse the samples may also influence the choice of isotope used. Both heavy oxygen and deuterium analyses are normally performed by isotope ratio mass spectrometry, whilst samples enriched with tritium are analysed by scintillation counting. Tritium analyses are normally cheaper than those for deuterium, and the equipment is available in more laboratories. On the basis of cost alone, the best approach would be to label large numbers of animals with tritium; however, tritium has the disadvantage that it is a radioactive isotope. Additional safety and legal considerations must therefore be taken into account before this isotope is used.

The last assumption when using the mark/recapture method is that one has delineated adequately the area over which the population is dispersed and that there are no hidden areas into which animals can disperse and exchange. Consider, for example, if one were attempting to measure the population of small mammals living on an island using mark and recapture. The estimated population size would be erroneous if, unknown to the person doing the study, the animals regularly swam between the island in question and another smaller island some distance offshore. In effect, one would be measuring the population of the two islands combined – leading to an overestimate of the population on the large island. This might appear to be a relatively minor problem for body composition analyses because individual animals are generally more easily defined than areas of habitat, unless the habitat is particularly fragmented, or one is dealing with island populations. Yet, there is one serious area where our assumptions about limits to the exchangeable population may be compromised and this is the problem of material inside the alimentary tract. Generally, material inside the alimentary tract is in isotopic exchange with the body as a whole. Hence, when one measures the isotope dilution space one is actually measuring the combined spaces of the body and the gut contents. Assuming all this water is in the body space will therefore lead to an overestimate – very much like in the small and large island example given above. This problem might be exacerbated even further if the isotopic label being used is deuterium or tritium, and the gut contains significant quantities of cellulose, because cellulose contains considerable amounts of exchangeable hydrogen. The differing quantities of exchangeable hydrogen in the gut is probably at least part of the reason underpinning the tremendous variation in the extent to which the hydrogen space exceeds the body water pool in different studies of the cow and other ruminants (Table 3.1), while studies of the carnivorous seals provide estimates in which the average overestimate is not only smaller (ranging from 1.6% to 4%) but also much more consistent across studies. Ideally then, starving animals (while allowing them continued access to water) to allow their guts to empty would potentially eliminate this source of variation in the estimated body water spaces by dilution – particularly if the label is for hydrogen. The practicalities of such starvation, however, will depend critically on the logistics of the given field situation and the amounts of stress that can realistically be imposed on wild animals without compromising other aspects of the study being performed. Calazel *et al.* (1993) found only minor impacts of fasting and feeding subjects during the equilibrium process in humans, but we will present data later in this chapter which indicate a significant effect in small animals.

There has been some debate over the nature of the exchangeable groups onto which hydrogen exchanges to produce the overestimated body water space. Several studies have suggested the source of the exchangeable hydrogen is amino groups on protein (Krough & Ussing, 1936–7; Ussing, 1938; Culebras & Moore, 1977). Another possibility is that label is exchanged with hydrogen groups on body lipids. This latter effect seems less likley since McClintock and Lifson (1957a, b) established that the excess dilution space in the Ob/Ob mouse is not abnormally elevated relative to lean mice (see Table 3.1). It seems most likely, therefore, that the major source of exchange is due to amino groups in body protein and exchangeable hydrogen on the gut contents, although the extent to which the hydrogen space exceeds the body water pool might be considered a problem for the isotope dilution techniques, and this is certainly the case when significant exchange occurs with respect to gut contents. It is perhaps important to consider whether the extent to which the hydrogen pool exceeds the body water pool (or the oxygen dilution space) can provide useful information on body composition. Three studies have examined the relation between the relative oxygen and hydrogen dilution spaces and body fatness in humans – two of which have produced no significant associations (Racette *et al.*, 1994; Goran *et al.*, 1992), but in the third a significant effect was detected (Westerterp *et al.*, 1995). To date, however, no studies have been published on animals. We have found significant positive relationships between body fatness and the extent to which the dilution space for hydrogen exceeds that for oxygen in both dogs and cats (J. R. Speakman, A. Hawthorne and R. Butterwick; unpublished data). These relationships to body fatness are perplexing, considering the suggestion that most of the excess hydrogen is probably exchanging with amino groups. However, it is often found that body lean mass and body fat mass are correlated, so a correlation may arise because of this association rather than any causal relationship. To date, we are aware of no studies that have examined links between the extent of difference in hydrogen and oxygen dilution spaces and individual variations in lean tissue content.

A final question concerning the use of the ratio of the hydrogen to the oxygen dilution spaces pertains to the problem of whether this ratio can be measured with sufficient precision to provide useful information on body composition. The problem is that the two spaces are both measured with error but differ by only a small amount (Table 3.1). If, for example, the estimate of each isotope dilution space had an accuracy of only 1% (s.d.), in combination the ratio of the two dilution spaces could be expected to vary by about 1.5% (ignoring any covariance in error). Hence an actual difference of 4% between spaces would have a 95% confidence limit from around +1 to +7%. This high individual variation in the ratio of the oxygen and hydrogen dilution spaces has been observed in almost all studies performed to date. The fact that this range exceeds the suggested physiological limits that are possible, given the amounts of exchangable hydrogen in the body (Culebras & Moore, 1977; Matthews & Gilker, 1995), suggests that no useful information can be gleaned from individual variations in this ratio.

It is even more surprising then that some studies have reported significant associations of the ratio to body fatness (above). One way of re-evaluating the question of precision in the ratio is to inject subjects with not only isotopes of ^{18}O and deuterium or tritium, but to include BOTH of the hydrogen isotopes along with the ^{18}O. If the analyses are then performed independently on both of the hydrogen isotopes, one would anticipate only a very weak correlation between the individual variation in dilution space ratios if most of the variability reflects analytical imprecision in the estimates for any of the isotopes. We are aware of only two studies which have performed this comparison with conflicting results. Arnould *et al.* (1996) compared the deuterium to the tritium dilution spaces in relation to the body water pool evaluated by dessication in four fur seals (*Arctocephalus gazella*) and found no correlation in the dilution space ratio (Fig. 3.4(*a*)). However, Król and Speakman (1999) found a very strong and significant link in the dilution space ratios established using deuterium and tritium (Fig. 3.4(*b*)) across 26 individual mice. This study strongly implies that individual dilution spaces lying outside the supposed physiological limits

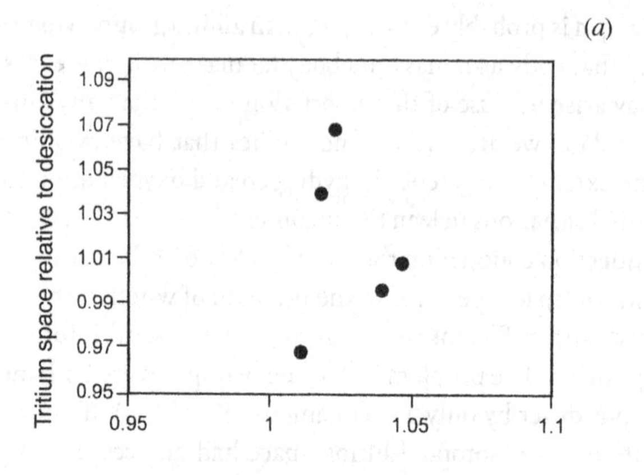

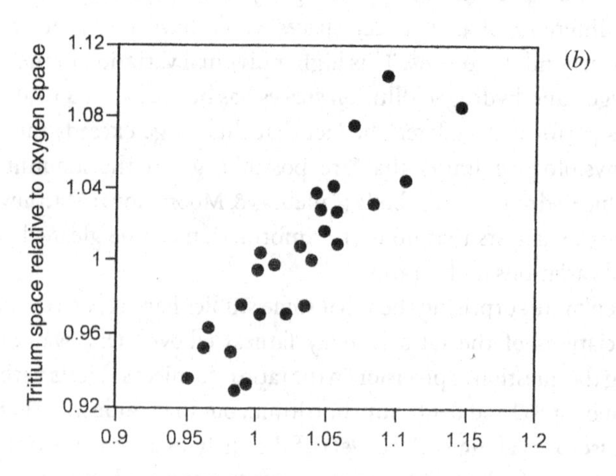

Fig. 3.4. Relationship between the hydrogen dilution spaces relative to body mass when using tritium and deuterium: (*a*) for four fur seals using data in Arnould *et al.* (1996), (*b*) for 26 mice using data in Król and Speakman (1999).

proposed by Culebras and Moore (1977) and Matthews and Gilker (1995) do not necessarily reflect analytical error, but more our ignorance about the processes of *in vivo* isotope exchange. More validation is certainly required to establish the extent to which the difference in dilution spaces of hydrogen and oxygen isotopes can provide valuable information on body composition.

Practical aspects

In the theoretical section we have described how the amount of total body water can be calculated from: (i) the quantity and isotope enrichment of the dose administered (M_{mol}, and E_{in}, respectively), (ii) the post-equilibrium isotope concentration in the body water pool (E_p), and (iii) the background concentration of the specific isotope in the animal's body water pool prior to administration of the label (E_b) (eqn. 4). For each dilution experiment, we have to optimize the values of these parameters to increase the overall accuracy and precision of the method, at a minimum of stress imposed on the animal, and at a minimum of costs for isotopes and analyses. In the following paragraphs we will review the sensitivity of each of these parameters to experimental errors, and we will give practical guidelines for the application of the isotope dilution method under laboratory and field conditions.

The dose administered

The isotope dose has to be administered quantitatively. From eqn. 4 it can be derived that there is a linear relationship between the error in the quantity of isotope which is administered and the overall error in the estimate of the total amount of body water. Thus, a 5% error in the determination of the amount of isotope administered results in a 5% error in the estimated body water. The amount of labelling isotope which is in a dose depends upon two factors: the enrichment of the isotope and the mass (or volume) of dose supplied.

The recommended dose of isotope depends upon the size of the subject and the accuracy and precision of the measurements of isotope enrichment in the samples (background and equilibrium) that are collected. Larger animals need larger doses to achieve an equivalent equilibrated isotopic enrichment, and procedures for analysis that result in more accurate and precise estimates of isotope enrichment may permit a lowering of the dose. We have modelled the resultant errors in the estimated size of the body water pool for a range of different equilibrium isotope enrichments above background, using a number of different levels of precision and accuracy in the isotope determinations. The results of this modelling are shown in Fig. 3.5. In general, there was an exponential decline in the error associated with the determination of the body water pool size as the plateau enrichment above background increases. For very low precision measurements of isotope enrichments

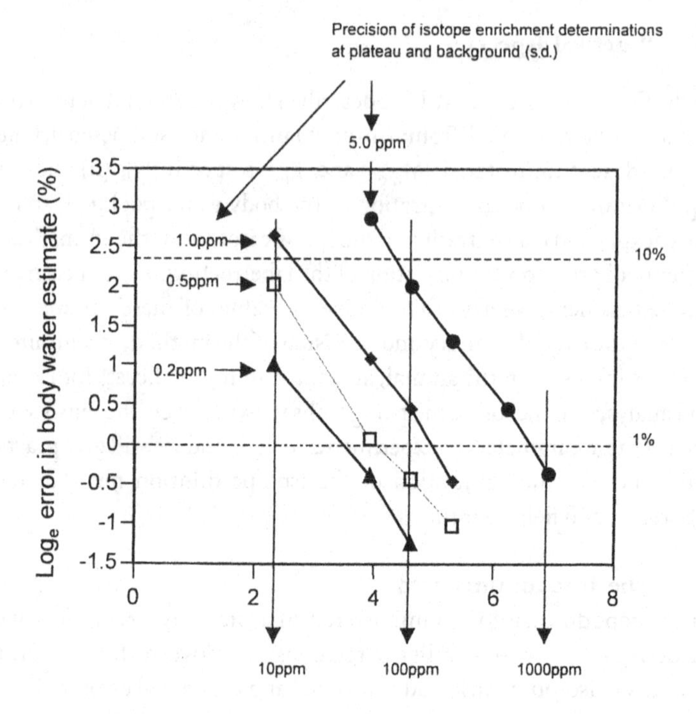

Fig. 3.5. The percentage error in the determination of body water space as a function of the extent to which isotope enrichment is elevated between the background and the equilibrium 'plateau' enrichment. Four lines are drawn, reflecting four different levels of precision in isotope enrichment determination. When precision is high (solid triangles), the degree of enrichment required to attain a certain level of error in the final body water estimate is lower than when the degree of precision in isotope determinations is low (solid circles).

(with an s.d. of 5 ppm), it is necessary to dose the animals with sufficient isotopes to raise the equilibrium enrichment to about 1000 ppm above background to obtain an estimate of body water that has a precision (s.d.) better than 1%. The critical increase in enrichment above background that is required to achieve 1% error (s.d.) in the body water determination is around 150 ppm when the error in isotope determination is only 1 ppm, 60 ppm when the error is 0.5 ppm and only 35 ppm when the error is 0.2 ppm (Fig. 3.5).

There is consequently a considerable saving in isotope costs to be made if procedures can be employed that minimize the error in the isotopic determination. For several reasons, however, the results of this modelling remain optimistic. In particular, it is assumed that the temporal location of the plateau (see Figs. 3.1–3.3) can be adequately established. In practice, therefore, an accuracy error may compromise these precision estimates. It is unlikely that dosages of isotopes for determination of body water content would ever reach levels at which analyses would be compromised because the enrichments were too high to be established by isotope ratio mass spectrometry. In general, it is probably best to aim for an elevation above background of between 50 and 200 ppm depending on the size of the animal, and the desire to minimize the costs of the dose.

Verification of the isotope concentration of the dose

Isotopes can be purchased from various sources, and are often delivered with a certificate listing the enrichments of the different isotopes therein. From eqn. 4, it can be shown that an error in the estimated isotope enrichment of the dose will result in an error in the estimate for the amount of body water in a linear manner. Therefore, it is recommended to make a dilution of the original source to verify its isotope concentrations (Halliday & Miller, 1977). The best way to do this is to mimic an injection of the dose into an organism's body water pool, by applying the same dose and the estimated average body water pool size of the study animal. Ideally, the isotope enrichments of these dilution samples should then be analysed alongside the experimental samples to remove machine effects (Prentice, 1990).

How to determine the quantity of the dose administered: weighing vs. volume readings

In many cases it is not possible to accurately weigh the syringe before and after administration of the isotopically labelled dose to the animal. This is especially the case for remote field locations, where it is not possible or desirable to transport the animal to a laboratory with an accurate balance. Under these circumstances, we have to estimate the amount of isotopes administered indirectly from a volume reading of a syringe. The following guidelines may be of help to minimize the error in the determination of the quantity of isotope administered.

(i) Use a syringe of appropriate size. For example, it is very difficult to inject 0.1 ml quantitatively when employing a 5 ml syringe. It is always best to use a syringe which is as small as possible, and with a scale as long as possible to enable an accurate reading. For example, the precision is much higher when using a 1 ml syringe with a 10 cm scale (a long syringe with small diameter) than one with a 5 cm scale (a short syringe with larger diameter).

(ii) Avoid the use of plastic syringes with a disposable needle. It appears to be very difficult to attach the needle to the syringe in a standard way. This results in considerable variation in the dead space between the plunger and the tip of the needle, and, thus, in a poor reproducibility of injected volumes between syringes. Much better results are obtained with medical glass syringes (Hamilton), or disposable insulin syringes of appropriate size with a fixed needle having a very small dead space. Medical glass syringes are available for a wide range of sizes (from 5 microlitre onwards), but the needles of these syringes are very long and easy to bend. Disposable insulin syringes of 0.3, 0.5, or 1.0 ml, with a fixed needle, have proved to be very useful in the field. For this type of syringe, errors in the injection volume are relatively small, partly due to the fact that there is only very small dead space volume between the plunger and needle. Another advantage of these syringes is that only very small quantities of isotopes remain in the dead space, which results in less waste of costly isotopes.

(iii) Calibrate each syringe in the laboratory by filling it with distilled water to the desired level. Weigh the syringe, empty it and reweigh again. Repeat this procedure several times for each syringe, using different volumes, which gives you an idea about the different 'errors' involved in the injection procedure. This is very helpful in designing the best injection procedure.

(iv) For animals with a small body water pool, volumes that need to be administered may become too small for accurate determination, resulting in the error likely to be larger than 1%. Dilution of the original isotope mixture allows administered volumes to be larger, which may result in a higher accuracy.

Storage of the dose

The isotope dilution to be administered can easily become diluted with atmospheric water vapour. Therefore, it is of extreme importance to avoid contact between the mixture and air by using a thick rubber stopper, and an aluminium cap. To reduce bacterial growth, the isotope mixture should be stored cool (5–10 °C). However, it has to be noted that freezing point of heavy water can be around 4 °C, depending on the concentration of the heavy isotopes. If the dose is allowed to freeze, the container may crack. The amount of moisture derived from the dose solution that would have a significant contamination effect on samples is minute (Schoeller *et al.*, 1995). It is always advisable, therefore, to locate the dose solution remote from samples derived from studies.

How to administer the dose

Ways to administer the isotopes include oral dosing as water, or injection using a syringe. Oral dosing is considered to be less stressful, and is typically applied in humans, and sometimes in larger animals. As a result of this route of administration, isotopes will enter in the stomach. From there, they will gradually distribute over the entire body water pool. In most other animals (especially the smaller ones), the dose is administered by injection into the animal's body water pool. In most cases, intravenous, subcutaneous, intramuscular (i.p.) or intraperitoneal (i.m.) injections are employed. In most studies on smaller animals, administration is performed by intraperitoneal injection. During injection, some water pressure will develop in the tissue or intraperitoneal cavity. If the needle is removed immediately after administration, there is a risk that some of the injectate may leave the body through the puncture. This may have an impact on the quantification of the dose that will mix with the body water pool. Therefore, injections should be performed slowly to avoid the building up of the water pressure. In addition, it is the best to keep the needle in its place for some seconds after injection, and use as small a gauge needle as possible.

Duration of equilibration period

Immediately after administration, the dose is situated in a specific place, e.g. stomach or intraperitoneal cavity, and needs to equilibrate with the entire body water pool (see Figs. 3.1–3.3). In an organism's body water pool, diffusion and cardiovascular circulation are the major mixing processes. It is not known after what exact time period equilibration will have been completed. Intuitively, one can imagine that equilibration time is shorter in smaller organisms than in larger, because the former exhibit a higher cardiac stroke frequency, and a smaller body water pool size, i.e. shorter distances resulting in shorter diffusion times to achieve equilibration.

In Fig. 3.6 we display five isotope equilibration curves for different-sized animals that were injected i.p. (blue tits, mice, Japanese quail), or i.m. (king penguins) or i.m. and i.v. (König horses). Similarities in the

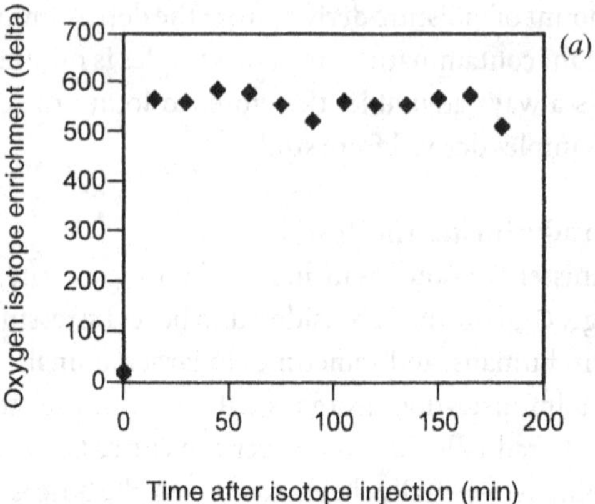

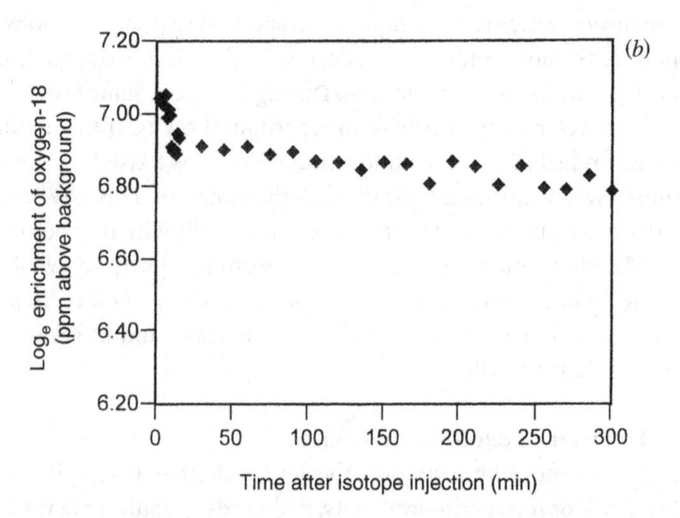

Fig 3.6. Empirical patterns of isotope enrichment following dosing with isotope of oxygen and hydrogen in animals of five different body sizes: (*a*) blue tits (13 g), oxygen-18, breath sampled (D.W. Thomas, J. Blondel and J.R. Speakman, unpublished data); (*b*) mice (25–30 g), oxygen-18, breath sampled (Król and Speakman, 1999), (*c*) Japanese quail (60–75 g), hydrogen, blood samples (G.H. Visser, unpublished data), (*d*) king penguins (3–6 kg), both isotopes, blood samples (G.H. Visser, unpublished data), (*e*) König horses (300 kg), hydrogen isotopes, i.v. and i.m. injection routes (G.H. Visser, unpublished data).

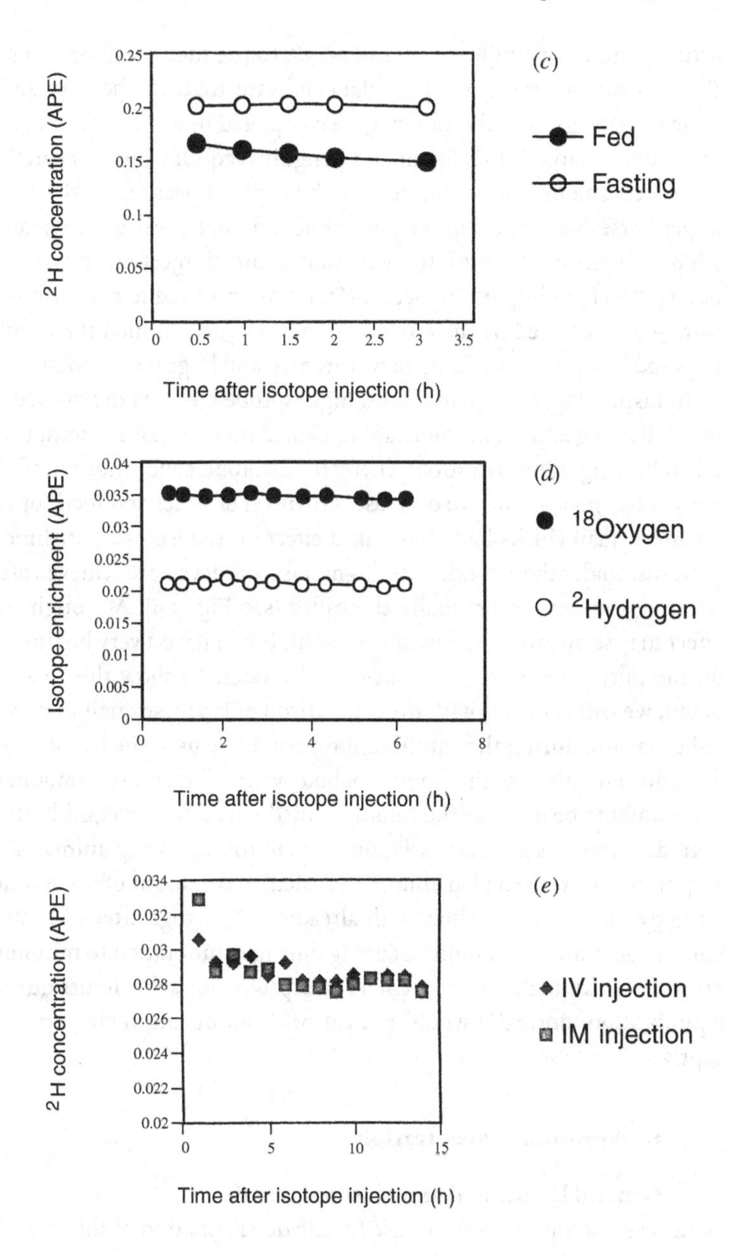
(c)
2 H concentration (APE)
0.25
0.2
0.15
0.1
0.05
0
0 0.5 1 1.5 2 2.5 3 3.5
Time after isotope injection (h)
Fed
Fasting
(d)
Isotope enrichment (APE)
0.04
0.035
0.03
0.025
0.02
0.015
0.01
0.005
0
0 2 4 6 8
Time after isotope injection (h)
18Oxygen
2Hydrogen
(e)
2 H concentration (APE)
0.034
0.032
0.03
0.028
0.026
0.024
0.022
0.02
0 5 10 15
Time after isotope injection (h)
IV injection
IM injection

actual patterns of equilibration in Fig. 3.6 to the theoretical patterns in Figs. 3.1–3.3 are evident. These data indicate that, in the very small animals (<30 g), equilibration may be completed in 15–20 minutes post-injection. In larger birds (quail and penguins) equilibrium appeared to have been attained by 30 minutes to 1 hour post-injection. In the much larger horse, however, equilibrium was not attained until approximately 7 h post-injection. In this latter situation, route of injection appeared to be a relatively unimportant factor on the time required to reach equilibrium – as suggested by the modelling in Fig. 3.3(*c*)/(*d*) when the limit is imposed by the rate of mixing between small and large body pools (k_m).

In fasting Japanese quail chicks, high isotope levels were achieved 30 min following administration, and remained more or less constant until 2 h following administration. Thereafter, isotope concentration of the body water pool seemed to decrease. Provision of water and food for the Japanese quail chicks had a profound effect on the isotope enrichment patterns. Under these conditions, highest isotope levels are achieved after 30 min, and decline gradually thereafter (see Fig. 3.6). Although this effect may seem fairly modest at first sight, it will have a very big impact on the calculation of the amount of body water. To show this in some detail, we will continue with the comparison of Japanese quail that were fed or fasting during the equilibration period. We used the isotope data and eqn. 4 to calculate the amount of body water. If we set the amount of body water to be 100% for the value obtained after a 30 min equilibration period, it can be seen that variation is small in the fasting animal (Fig. 3.7). In contrast, in the fed animal the calculated amount of body water strongly increases with time, with already a 4% change after 1 h. These data suggest animals should be fasting during equilibration to minimise errors. Wild animals, which are held in captivity for a few hours during equilibration, normally would not eat or drink during their period in captivity.

Background concentration

General latitudinal and seasonal trends

Whichever isotope we want to use for the determination of the animal's body water pool, we need to know the concentration of that isotope in the animal's body water pool prior to administration (E_b, eqn. 4). In seawater concentrations of ^2H, ^3H, ^{17}O, and ^{18}O are about 0.015%, <10–15%, 0.0035%, and 0.200%, respectively. However, in all organisms, these back-

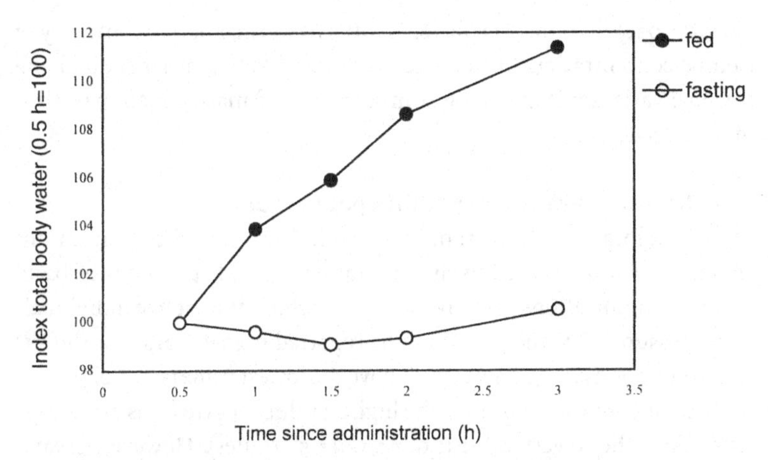

Fig. 3.7. Changes in the estimated body water content form isotope enrichment estimates at different times following injection for Japanese quail (Fig. 3.6(*c*)). All estimates are expressed relative to the pool space at time 0.5 h post-injection. For animals allowed to feed, there were large changes in the pool size estimates over time.

ground concentrations vary depending on the isotopic composition of the drinking water and water attached to the food, as well as the relative levels of evaporative water loss from the animal's body water pool, due to the process of fractionation. In general, animals living in arid environments have relatively high background concentrations, whereas they are relatively low in animals inhabiting temperate and polar terrestrial habitats. This pattern more or less follows the general relationship between the isotope enrichment in freshwater reservoirs and latitude (Dansgaard, 1964; Taylor, 1974). For example, in Kentish plovers (*Charadrius alexandrinus*) breeding in southern Spain, ^{2}H and ^{18}O background concentrations are 153.7 and 2004.5 ppm, respectively (Amat *et al.*, 2000). However, for the high arctic breeding ringed plover (*Charadrius hiaticula*), the corresponding values are 139.3 and 1978.3 ppm, respectively (T. Piersma, unpublished data). The latitudinal trend is much stronger for ^{2}H (factor: 1.103) than for ^{18}O (factor: 1.013). In addition to this general latitudinal trend, there is a seasonal trend that isotope background levels tend to be higher during summer than during the winter (Tatner, 1990). For example, in the European robin (*Erithacus rubecula*) ^{2}H and ^{18}O background concentrations in Scotland in July are 153.1 and 2009.6 ppm, respectively, but in November 145.6 and 1990.6 ppm, respectively. As observed for the latitudinal trend, the variation is relatively larger for ^{2}H

(factor: 1.052) than for ^{18}O (factor 1.001). Given the relative constancy of isotope concentrations in seawater, it is very likely that latitudinal and seasonal variation is much lower in seabirds and marine mammals than in terrestrial animals.

Individual variation within a population

At a given location and season, variation in background concentrations among individuals is relatively low. Tatner (1990) lists ranges of background concentrations from five studies in which at least five individuals were measured. For these studies, the reported highest value for the ^{18}O isotope was on average only 0.73% above the lowest value (s.d. = 0.41, $n = 5$ studies), and for the ^{2}H isotope the highest value reported was on average 2.3% above the lowest value (s.d. = 2.1, $n = 5$ studies). However, greater levels of variation have been detected in some species, for example, ptarmigan (Thomas *et al.*, 1994).

Do we need to determine the background concentration for each population or individual?

After having reviewed the general latitudinal and seasonal patterns in isotope background levels, one may start wondering whether it is really necessary to determine the individual-specific background level or whether a general population-specific estimate would suffice. The latter option would be advantageous, since this protocol is less stressful for the study animals and less samples are generated for analysis. If we have a closer look at eqn. 4, it can be seen that E_b only appears in the denominator, in which the difference between the plateau enrichment (E_p) and the background enrichment (E_b) is calculated. If there is a large difference between E_p and E_b after administration of the label, e.g. due to a relatively high dose, any uncertainty with respect the exact background level will have little impact on the overall estimate for the amount of body water. In contrast, if the difference between E_p and E_b is small (relatively low dose), any variation in E_b will have a considerable effect on the estimate of the amount of body water. So, at a relatively high dose, for most species, there is no need to determine the individual-specific background level. We modelled this effect more closely by setting the level of variation in the background isotopes at the level established by Tatner (1990), i.e. with a range of 0.73% for the oxygen isotope and 2.3% for the hydrogen isotope. At the given background levels of around 2000 ppm for ^{18}O and 150 ppm for deuterium, these translate to ranges of 14.6 ppm and 3.45 ppm, respectively. If we assume the s.d. for the interindividual variation is about half the range, the resultant error can be estimated for different doses and various precisions of determining the plateau enrichment. This modelling revealed, as anticipated, that the dose of isotope required to achieve a 1% error in the estimated body water pool size was much higher than when a background sample was taken, except when the precision of isotope determination was around the same level as the variation between individuals in the background enrichment. In this situation, there is no virtue in taking a background sample because it does not improve the precision of the resultant estimate and the dose required remains the same. A less obvious result, however, was that because the error in the estimated body water becomes limited by the precision of the background estimate, the dose required to achieve 1% error in the body water was far less dependent on the precision of the isotope determination of the plateau enrichment. Thus, for oxygen the increase in enrichment required to achieve 1% error was 650, 800, 900 and 1000 ppm for precisions of determining the plateau enrichment of 0.2, 0.5, 1.0 and 5.0 ppm, respectively, when no background sample was taken, compared with increases of 35, 60, 150 and 1000 ppm when a background was taken pre-injection.

This modelling has some important implications for the decision of whether to take a background sample or not. First, if the precision of isotope determination is poor (s.d. = 5 ppm or worse) there is no improvement in the precision of the final body water estimate by taking an individual background sample. Secondly, if the precision of the isotope determinations is good (s.d. = 1 ppm or better) then the dosage of isotopes can be reduced by a factor of between 5 and 20 fold if an individual background sample is taken prior to injection. The cost of the dose is likely to become an increasingly important issue as the size of the animal increases, but the difficulty of multiple sampling becomes an issue as the size of the animal decline. It seems most likely, therefore, that the optimal protocols for larger animals will involve low doses and taking of a background sample, while for smaller animals the optimal protocol might involve not taking a background sample and using larger doses. The trade-off between these approaches is likely to be defined by the volume requirements for isotope determinations. Hence using pyrolysis procedures it is possible to obtain isotope determinations for deuterium which only require samples of 1–2 μl. Multiple sampling of this volume will be possible from considerably smaller animals than would be possible if ^{18}O was the label and the analysis procedure involved isotope equilibration, where optimally in excess of 50 μl of body water is used.

The samples

Which type of sample to take

The body water pool can be sampled in several different ways. The most direct method is undoubtedly to take a blood sample. This has the benefit of being instantaneously drawn from the pool of interest. Saliva provides another amenable source of water as does urine. It is important to note that saliva samples could be compromised if the dose has been administered orally because the oral cavity may contain residual traces of the dose – compromising the isotope determination. Urine also poses problems because it is a reservoir of body water that integrates incoming water from the body water pool over an unknown time period. Perhaps the only animals from which reliable urine samples can be collected are humans – where it is possible to instruct subjects to void their bladders completely and then after a short period to allow further urine to accumulate to void the sample. It is possible that animals could be induced to comply with such a protocol, by forcing them to urinate on multiple occasions but this is always likley to be less efficient than taking blood. Gotaas *et al.* (1997) recovered droppings produced by animals in the wild and extracted water from them to estimate the isotope enrichments in body water without the need to trap the animals. We have recently used a similar method to collect urinations produced by meercats (M. Scantlebury, T.H. Clutton-Brock and J.R. Speakman, unpublished data), although comparisons to blood collected simultaneously suggest this method does not always provide reliable estimates of body water content. Król and Speakman (1999) collected samples of CO_2 in the expired breath of mice to evaluate the ^{18}O enrichment of their body water by confining them in a small chamber to which a needle valve had been attached, onto which a vacutainer could be inserted. This provided a completley non-invasive method for estimating the body isotope enrichment and, as such, it was possible to take multiple samples at frequent intervals (every minute) to establish the time course of isotope equilibration in this small animal (Fig. 3.6(*b*)). Such rapid and frequent sampling would not be possible in any other way. For lactating females another source of body water is expressed milk – although Scantlebury *et al.* (2000) found significant discrepancies between milk enrichments of deuterium and those simultaneously determined in blood, perhaps due to exchangeable deuterium previously incoprorated into lipids also secreted into the milk.

Two studies have made comparisons of the estimated enrichments of body water drawn from a variety of sources during the process of equilibration (Wong *et al.*, 1988; Schoeller *et al.*, 1986). Both suggested that blood, saliva, breath and urine provided effectively comparable estimates of the body water enrichment once the plateau had been achieved, but there were some discrepancies attributable to differential mixing in the early stages of the equilibrium process.

How to store the samples

Isotopically enriched and background samples from the subject must not alter in their isotopic enrichment during storage. The lighter isotope from isotopically enriched water samples will evaporate more quickly than the heavy isotopes, so it is important that samples are stored in sealed containers as quickly as possible. Large samples (several ml) such as urine from humans and large animals, can be stored in vials with screw caps which are then frozen. Westerterp *et al.* (1995) demonstrated that samples stored in plastic vials lose their enrichment over time and glass is the recommended storage vessel. Blood samples from small animals (10–100 μl) are typically stored inside flame-sealed glass capillaries. The capillaries should be kept cool to reduce possible bacterial growth in the sample, but they must not be frozen since that would result in the glass cracking. As little additional air as possible should be included with the sample, given the constraints that samples in vials must be allowed room for expansion if they are to be frozen, and the sample must not be heated during flame sealing of glass capillaries. Breath samples can be collected into vacutainers and stored for protracted periods prior to analysis (K. Hatch *et al.*, unpublished data).

Options for isotope analysis

Tritium is analysed by scintillation counting. Deuterium can be analysed by infrared spectroscopy (Stanstell & Mojica, 1968), but the doses required to obtain a reasonable measurement are high, increasing the costs of dosing, and the precision of the resultant estimates is poor. Currently the most effective method for determining the isotope enrichments of deuterium and ^{18}O or ^{17}O in the range from background to +500 ppm is gas source isotope ratio mass spectrometry (IRMS). In this enrichment range precision is routinely better than 0.5 ppm and precisions of 0.1 ppm can be achieved with some preparation processes.

For oxygen analyses by IRMS the most frequently used preparation method is equilibration with CO_2. This procedure requires a sample of at least 10 μl, and ideally a sample exceeding 50 μl is required (Speakman *et al.*, 1990). If samples lower than 10 μl need to be analysed, the most suitable methods are the guanidine conversion procedure (Boyer *et al.*, 1961; Dugan *et al.*, 1985). An alternative approach is by cyclotron irradiation and counting of the decay products of transformed nuclei (Amiel & Piesach, 1963; Wood *et al.*, 1975).

For deuterium analysis, the most commonly used methods include reduction of the water by metals heated to high temperatures. Wong and Klein (1987) reviewed eight different procedures in use at the time. Since then, several other methods have come into popular use. These include reduction by lithium aluminium hydride (Ward *et al.*, 2000), on-line pyrolysis (Begley & Scrimgeour, 1996, 1997) and conversion to acetylene using calcium carbide (van Kreel *et al.*, 1996). All these methods work effectively with samples between 5 and 20 μl. A recent and exciting advance in isotope determination methodology is the successful development of laser spectroscopy (Kerstel *et al.*, 1999) . This provides a rapid method for determination of deuterium, ^{17}O and ^{18}O levels simultaneously in small (20 μl) blood samples without the need for prior preparation by distillation.

Conversion of isotope dilution space to the amounts of body water, dry lean mass and fat mass

In most studies, the determination of the amount of body water is not a goal in itself, but it is a route to indirectly estimate the amounts of lean matter and fat in the animal's body. This conversion is possible because lean and fat tissue differ in their water contents. Fat is considerably more anhydrous than lean tissue. Hence as an animal gets progessively fatter the percentage of its total body mass that comprises water becomes progressively reduced. It is generally considered that lean tissue containing only structural fat contains approximatley 73% water (Pace & Rathbun, 1945; Fuller *et al.*, 1992). This estimate has been validated on many different occasions, and a theoretical treatment concerning why the value should be relatively stable at this level has been developed. However, while the average value appears relatively stable, individual variation can be large. Moreover, there are significant population deviations from this value. For example, young animals have lean tissues that are considerably wetter, e.g. Sawicka-Kapusta, 1974.

The absence of any absolute value for the water content of lean tissue is perhaps the most significant problem for the conversion of body water estimates derived from isotope dilution into estimates of total body lean and fat contents. Inevitably, to minimize error in this conversion, it is desirable to perform some calibration work by comparing the isotope dilution estimates of body water to direct estimates of lean and fat tissue contents (see Hildebrand *et al.*, 1998 for an example of such a calibration exercise and see Chapter 2, for details of destructive methods of analysis).

REFERENCES

Amat, J.A., Visser, G.H., Perez-Hurtado, A. & Arroyo, G.M. (2000). Brood desertion by female shorebirds: a test of the differential parental capacity hypothesis on Kentish plovers. *Proceedings of the Royal Society of London* B, **267**, 2171–6.

Amiel, S. & Piesach, M. (1963). Oxygen-18 determination by counting delayed neutrons of nitrogen-17. *Analytical Chemistry*, **35**, 323–7.

Andrew, S.M., Erdman, R.A. & Waldo, D.R. (1995). Prediction of body composition of dairy cows at three physiological stages from deuterium oxide and urea dilution. *Journal of Dairy Science*, **78**, 1083–95.

Arnold, R.N., Hentges, E.J. & Trenkle, A. (1985). Evaluation of the use of deueterium-oxide dilution techniques for determination of body-composition of beef steers. *Journal of Animal Science*, **60**, 1188–200.

Arnould, J.P.Y., Boyd, I.L. & Speakman, J.R. (1996). Measuring the body composition of Antarctic fur seals (*Arctocephalus gazella*): validation of hydrogen isotope dilution. *Physiological Zoology*, **69**, 93–116.

Barratt, E.M., Deaville, R., Burland, T.M. *et al.* (1997). DNA answers the call of pipistrelle bat species. *Nature*, **387** (6629), 138–9.

Begley, I.S & Scrimgeour, C.M. (1996). On-line reduction of H_2O for delta H-2 and delta O-18 measurement by continuous-flow isotope ratio mass spectrometry. *Rapid Communications in Mass Spectrometry*, **10** (8), 969–73.

Begley, I.S. & Scrimgeour, C.M. (1997). High-precision delta H-2 and delta O-18 measurement for water and volatile organic compounds by continuous-flow pyrolysis isotope ratio mass spectrometry. *Analytical Chemistry*, **69** (8), 1530–5.

Bell, G.E. & Stern, J.S. (1977). Evaluation of body composition of young obese and lean Zucker rats. *Growth*, **41**, 63.

Bohm, B.C. & Hadley, N.F. (1977). Tritium determined water flux in the free roaming tenebrionid beetle, *Eleodes armata*. *Ecology*, **58**, 407–14.

Boyer, P.D., Graves, D.J., Suelter, C.H. & Dempsey, M.E. (1961). Simple procedure for conversion of oxygen of orthophosphate or water to carbon dioxide for oxygen-18 determination. *Analytical Chemistry*, **33**, 1906–9.

Calazel, C.M., Young, V.R., Evans, W.J. & Roberts, S.B. (1993). Effect of fasting and feeding on measurement of CO_2 production using doubly-labeled water. *Journal of Applied Physiology*, **74**, 1824–9.

Carnegie, A.B. & Tulloch, N.M. (1968). The *in vivo* determination of body water space in cattle using the tritium dilution technique. *Proceedings of the Australian Society of Animal Production*, **7**, 308–13.

Chinard, F.P. & Enns, T. (1954). Transcapillary pulmonary exchange of water in the dog. *American Journal of Physiology*, **178**, 197–203.

Crabtree, R.M., Houseman, R.A. & Kay, M. (1974). The estimation of body composition in beef cattle by deuterium oxide dilution. *Proceedings of the Nutrition Society*, **33**, 74A–75A.

Crum, B.G., Williams, J.B. & Nagy, K.A. (1985). Can tritiated water dilution space accurately predict total body water in Chukar partridges? *Journal of Applied Physiology*, **59**, 1383–8.

Culebras, J.M. & Moore, F.D. (1977). Total body water and the exchangeable hydrogen I. Theoretical calculation of nonaqueous exchangeable hydrogen in man. *American Journal of Physiology*, **232**, R54–9.

Culebras, J.M., Fitzpatrock, G.F., Brennan, M.F. *et al.* (1977). Total body water and the exchangeable hydrogen. II. A review of comparative data from animals based on isotope dilution and desiccation, with a report of new data from the rat. *American Journal of Physiology*, **232**, R60–5.

Dansgaard, N. (1964). Stable isotopes in precipitation. *Tellus*, **16**, 436–68.

Degen, A.A., Pinshow, B., Alkon, P.U. & Arnon, H. (1981). Tritiated water for estimating total body water and water turnover rate in birds. *Journal of Applied Physiology*, **51**, 1183–8.

Dugan, J.P., Borthwick, J., Harmon, R.S. *et al.* (1985). Guanidine hydrochloride method for determination of water oxygen isotope and O-18 fractionation between carbon dioxide and water at 25 °C. *Analytical Chemistry*, **57**, 1734–6.

Edelman, I.S. (1952). Exchange of water between blood and tissues. Characteristics of deuterium oxide equilibration in body water. *American Journal of Physiology*, **171**, 279–96.

Eidinof, M.L., Perri, G.C., Knoll, J.E., Marano, B.J. & Arnheim, J. (1953). The fractionation of hydrogen isotopes in biological systems. *Journal of American Chemical Society*, **75**, 248–9.

Ellis, W.A.H, Marples, T.G. & Phillips, W.R. (1991). The effects of a temperature-determined food supply on the annual activity cycle of the lesser long-eared bat, *Nyctophilus geoffroyi* Leach, 1821 (Microchiroptera, Vespertilionidae). *Australian Journal of Zoology*, **39** (3), 263–71.

Enns, T. & Chinard, F.P. (1956). Relative rates of passage of $H^1H^3O^{16}$ and of $H^1{}_2O^{18}$ across pulmonary capillary vessels in the dog. *American Journal of Physiology*, **185**, 133–6.

Farrell, D.J. (1974). Effects of dietary energy concentration on utilisation of energy by broiler chickens and on body composition determined by carcass analysis and predicted using tritium. *British Poultry Science*, **15**, 25–41.

Farrell, D.J. & Belnave, D. (1977). The *in vivo* estimation of body fat content in laying hens. *British Poultry Science*, **18**, 381–4

Farrell, D.J. & Reardon, T.J. (1972). Undernutrition in grazing sheep. III. Body composition and its estimation *in vivo*. *Australian Journal of Agricultural Research*, **23**, 511–17.

Ferrell, C.L & Cornelius, S.G. (1984). Estimation of body composition of pigs. *Journal of Animal Science*, **58** (4), 903–12

Flynn, M.A., Hanna, F., Long, C.H. *et al.* (1968). Deuterium-oxide dilution as a predictor of body composition in children and pigs. In *Body Composition in Animals and Man*, ed. J. Brozek. Washington DC National Academy of Sciences, pub no. 1598, pp. 480–91.

Foot, J.Z., Skedd, E. & McFarlane, D.N. (1979). Body composition in lactating sheep and its indirect measurement in the live animal using tritiated water. *Journal of Agricultural Science Cambridge*, **92**, 69–81.

Foot, J.Z. & Greenhalgh, J.F.D. (1970). The use of deuterium oxide space to determine the amount of body fat in pregnant blackface ewes. *British Journal of Nutrition*, **24**, 815.

Foy, J.M. & Schneiden, H. (1960). Estimation of total body water (virtual tritium space) in the rat, cat, rabbit, guinea-pig and man, and of the biological half life of tritium in man. *Journal of Physiolology*, **154**, 169–76.

Freis, E.D., Higgins, T.F. & Morowitz, H.J. (1953). Transcapillary exchange rates of deuterium oxide and thiocyanate in the forearm of man. *Journal of Applied Physiology*, **5**, 526–32.

Fuller, N.J., Jebb, S.A., Laskey, M.A., Coward, W.A. & Elia, M. (1992). Four-component model for the assessment of body composition in humans: comparison with alternative methods, and evaluation of the density and hydration of fat-free mass. *Clinical Science*, **82**, 687–93.

Gettinger, R.D. (1983). Use of doubly labelled water ($^3H_2{}^{18}O$) for determination of H_2O flux and CO_2 production by a mammal in a humid environment. *Oecologia*, **59**, 54–7.

Glascock, R.F. & Duncombe, W.G. (1952). Biological fractionation of hydrogen isotopes in mammary gland and other tissues. *Biochemical Journal*, **51**, xl.

Goran, M.I., Poehlman, E.T., Nair, K.S. & Danforth, E.J.R. (1992). Effect of gender, body composition, and equilibration time on the ^2H-to-^{18}O dilution space ratio. *American Journal of Physiology* **263**, E1119–24.

Gordon, A.J., Topps, J.H. & Begg, T.W. (1971). Total body water of rats as measured with different amounts of injected tritiated water. *Proceedings of the Nutrition Society*, **30**, 55A.

Gotaas, G., Milne, E., Haggarty, P. & Tyler, N.J.C. (1997). Use of feces to estimate isotopic abundance in doubly labeled water studies in reindeer in summer and winter. *American Journal of Physiology – Regulatory Integrative and Comparative Physiology*, **42**, R1451–6.

Groves, T.D.D. & Wood, A.J. (1965). Body composition studies on the suckling pig. II The *in vivo* determination of body water. *Canadian Journal of Animal Science*, **45**, 14.

Halliday, D. & Miller, A.G. (1977). Precise measurement of total body water using trace quantities of deuterium oxide. *Biomedical Mass Spectrometry*, 4, 82–7.

Hevesy, G. von & Hofer, E. (1934). Diplogen in fish. *Nature*, **133**, 797–8.

Hevesy, G. von & Jacobson, C.F. (1940). Rate of passage of water through capillary and cell walls. *Acta Physiologica Scandinavica*, **1**, 11–18.

Hildebrand, G.V., Farley, S.D. & Robbins, C.T. (1998). Predicting body condition of bears via two field methods. *Journal of Wildlife Management*, **62**, 406–9.

Hollander, V., Chang, P. & Co Tui, F.W. (1949). Deuterium oxide and thiocyanate spaces in protein depletion. *Journal of Laboratory and Clinical Medicine*, **34**, 680–7.

Holleman, D.F. & Dietrich, R.A. (1973). Body water content and turnover in several species of rodents as evaluated by the tritiated water method. *Journal of Mammalogy*, **54**, 456–65.

Holleman, D.F. & Dietrich, R.A. (1975). An evaluation of the tritiated water method for estimating body water in small rodents. *Canadian Journal of Zoology*, **53**, 1376–8.

Houseman, R.A., McDonald, I. & Pennie, K. (1973). The measurement of total body water in living pigs by deuterium oxide dilution and its relation to body water composition. *British Journal of Nutrition*, **30**, 149–56.

Johnson, R.J. & Farrell, D.J. (1988). The prediction of body composition in poultry by estimation *in vivo* of total body water with tritiated water and deuterium oxide. *British Journal of Nutrition*, **59**, 109–24.

Jones, G. & Parijs, S.M. van (1993). Bimodal echolocation in pipistrelle bats – are cryptic species present. *Proceedings of the Royal Society of London, Series B*, **251** (1331), 119–25.

Jungas, R.L. (1968). Fatty acid synthesis in adipose tissues incubated in tritiated water. *Biochemistry*, **7**, 3708–17.

Karasov, W.H. (1983). Wintertime energy conservation by huddling in Antelope ground squirrels (*Ammospermophilus leucurus*). *Journal of Mammalogy*, **64**, 341–5.

Kay, M., Jones, A.S. & Smart, R. (1966). The use of tritiated water, 4-aminoantipyrene and N-acetyl-4-aminoantipyrene for the measurement of body water in living pigs. *British Journal of Nutrition*, **20**, 439–48.

Kerstel, E.R.T., Trigt, R. van, Dam, N. van, Reuss, J. & Meijer H.A.J. (1999). Simultaneous determination of the H-2/H-1, O-17/O-16, and O-18/O-16 isotope abundance ratios in water by means of laser spectrometry. *Analytical Chemistry*, **71** (23), 5297–303.

King, W.W. (1976). Energy metabolism and body water turnover rates of free-living scorpions. PhD dissertation, p. 104. Tempe, AZ: Arizona State University.

Kirchgessner, M., Roth-Maier, D.A., Kirch, P. & Schmidt, H.L. (1977). Zur Bestimmung des Gesamtkörperwassers Lebender Hühner mit der D_2O-Verdünnungsmethode (Estimation of total body water in the growing domestic fowl by the deuterium oxide dilution technique). *Zeitschrift für Tierphysiologie, Tierernahrung und Futtermittelkunde*, **39**, 104–8.

Kirshenbaum, I. (1951). In *Physical Properties and Analysis of Heavy Water*, ed. H. Urey & G.M. Murphy, pp. 1–41 and 187–259. New York: McGraw-Hill.

Kodama, A.M. (1971). *In vivo* and *in vitro* determinations of body fat and body water in the hamster. *Journal of Applied Physiology*, **31**, 218–22.

Kreel, B.K van, Van der Vegt, F., Meers, M. *et al.* (1996). Determination of total body water by a simple and rapid mass spectrometric method. *Journal of Mass Spectrometry*, **31**, 108–11.

Król, E. & Speakman, J.R. (1999). Isotope dilution spaces of mice injected simultaneously with deuterium, tritium and oxygen-18. *The Journal of Experimental Biology*, **202**, 2839–49.

Krough, A. & Ussing, H.H. (1936–37). The exchange of hydrogen between free water and the organic substances in the living organism. *Scandinavian Archives of Physiology*, **75**, 90–104.

Lifson, N., Gordon, G.B. & McClintock, R. (1955). Measurement of total carbon dioxide production by means of $D_2{}^{18}O$. *Journal of Applied Physiology*, **7**, 704–10.

Little, D.J. & McLean, R.W. (1981). Estimation of the body chemical composition in live cattle varying widely in fat content. *Journal of Agriculture Science*, **96**, 213–20.

Lucke, B. & Harvey, E.N. (1935). The permeability of living cells to heavy water (deuterium oxide). *Journal of Cellular Comparative Physiology*, **5**, 473–82.

Lydersen, C., Griffiths, D., Gjertz, I. & Wiig, O. (1992). A tritiated water experiment on a male atlantic walrus (*Odobenus rosmarus rosmarus*). *Marine Mammal Science*, **8**, 418–20.

McClintock, R. & Lifson, N. (1957a). Applicability of the D_2O^{18} method to the measurement of the total carbon dioxide output of obese mice. *Journal of Biological Chemistry*, **226**, 153–6.

McClintock, R. & Lifson, N. (1957b). CO_2 output and energy balance of hereditary obese mice. *American Journal of Physiology*, **189**(3), 463–9.

McClintock, R. & Lifson, N. (1958). Determination of the total carbon dioxide outputs of rats by the D_2O^{18} method. *American Journal of Physiology*, **192**(1), 76–8.

Matthews, D.E. & Gilker, C.D. (1995). Impact of ^{2}H and ^{18}O pool size determinations on the calculations of total energy expenditure. *Obesity Research*, **3**, 21–9.

Meissner, H.H. (1976). Urea space versus tritiated water space as an *in vivo* predictor of body water and body fat. *South African Journal of Animal Science*, **6**, 171–8.

Meissner, H.H., Staden, J.H. van & Pretorius, E. (1980). *In vivo* estimation of body composition in cattle with tritium and urea dilution. II. Accuracy of prediction equations of the chemical analyzed carcass and non-carcass components. *South African Journal of Animal Science*, **10**, 175–81.

Moore, F.D. (1946). Detemination of total body water and solids with isotopes. *Science*, **2**, 245.

Nagy, K.A. & Costa, D.P. (1980). Water flux in animals: analysis of potential errors in the tritiated water method. *American Journal of Physiology*, **238**, R454–65.

Nagy, K.A., Seymour, R.S., Lee, A.K. & Braithwaite, R. (1978). Energy and water budgets of free-living *Antechinus stuartii* (Marsupialia: Dasyuridae). *Journal of Mammalogy*, **59**, 60–8.

Odwongo, W.O., Conrad, H.R. & Staubus A.E. (1984). The use of deuterium oxide for the prediction of body composition in live dairy cattle. *Journal of Nutrition*, **114**, 2127–37.

Oftedal, O.T., Bowen, W.D. & Boness, D.J. (1996). Lactation performance and nutrient deposition in pups of the harp seal *Phoca groenlandica* on ice floes off south east Labrador. *Physiological Zoology*, **69**, 635–57.

Pace, N. & Rathbun, E.N. (1945). Studies on body composition. III. The body water and chemically combined nitrogen content in relation to fat content. *Journal of Biological Chemistry*, **158**, 685–91.

Pace, N., Kline, L., Schachman, H.K. & Harfenist, M. (1947). Studies on body composition. IV. The use of radioactive hydrogen for measurement *in vivo* of total body water. *Journal of Biological Chemistry*, **168**, 459–69.

Panaretto, B.A. (1963a). Body composition *in vivo* I. The estimation of total body water with antipyrine and the relation of total body water to total body fat in rabbits. *Australian Journal of Agriculture Research*, **14**, 594–601.

Panaretto, B.A. (1963b). Body composition *in vivo*. III. The composition of living ruminants and its relation to the tritiated water spaces. *Australian Journal of Agriculture Research*, **14**, 944–52.

Panaretto, B.A. (1968). Body composition *in vivo*. IX. The relation of body composition to the tritiated water spaces of ewes and wethers fasted for short periods. *Australian Journal of Agriculture Research*, **19**, 267–72.

Panaretto, B.A. & Till, A.R. (1963). Body composition *in vivo*. II. The composition of mature goats and its relationship to the antipyrine, tritiated water and N-acetyl-4-aminoantipyrene spaces. *Australian Journal of Agriculture Research*, **14**, 926–43.

Prentice, A.M. (ed.) (1990). *The Doubly-Labelled Water Method for Measuring Energy Expenditure: Technical recommendations for use in humans*. A consensus report by the IDECG Working Group, International Atomic Energy Agency, NAHRES-4, Vienna.

Racette, S.B., Schoeller, D.A., Luke, A.H., Shay, K., Hnilicka, J. & Kushner, R.F. (1994). Relative dilution spaces of ^{2}H- and ^{18}O-labelled water in humans. *American Journal of Physiology*, **267**, E585–90.

Reid, J.T., Balch, C.C. & Glascock, R.F. (1958). The use of tritium, of antipyrine and of N-acetyl-4-aminoantipyrene in the measurement of body water in living rabbits. *British Journal of Nutrition*, **12**, 43–51.

Reilly, J.J. & Fedak, M.A. (1990). Measurement of body composition in living gray seals by hydrogen isotope dilution. *Journal of Applied Physiology*, **69**, 885–91.

Rumpler, W.V., Allen, M.E., Ullrey, D.E., Earle, R.D., Schmidt, S.M. & Cooley, T.M. (1987). Body composition of white-tailed deer estimated by deuterium oxide dilution. *Canadian Journal of Zoology*, **65**, 204–8.

Sawicka-Kapusta, K. (1974). Changes in the gross body composition and energy value of the bank voles during their postnatal development. *Acta Theriologica*, **19** (3), 27–54.

Scantlebury, M., Hynds, W., Booles, D. & Speakman, J.R. (2000). Isotope recycling in lactating dogs (*Canis familiaris*). *American Journal of Physiology*, **278**, R669–76.

Schoeller, D.A., Leitch, C.A. & Brown, C. (1986). Doubly labeled water method: *In vivo* oxygen and hydrogen isotope fractionation. *American Journal of Physiology*, **251**, R1137–43.

Schoeller D.A., Taylor, P.D. & Shay, K. (1995). Analytic requirements for the doubly labeled water method. *Obesity Research*, **3**, 15–20.

Searle, T.W. (1970). Body composition in lambs and young sheep and its prediction *in vivo* from tritiated water space and body weight. *Journal of Agricultural Science, Cambridge*, **82**, 105–22.

Seber, G.A.F. (1982). *The Estimation of Animal Abundance*. 2nd edn. London:Griffin.

Setiabudi, M., Kamonsakpithak, S., Sheng, H.P. & Huggins, R.A. (1975). Growth of the pigs: changes in body weight and body fluid compartments. *Growth*, **40**, 405.

Sheng, H.P. & Huggins, R.A. (1971). Direct and indirect measurement of total body water in the growing beagle. *Proceedings of the Society for Experimental Biology and Medicine*, **137**, 1093–9.

Shields, R.G., Mahan, D.C. & Byers, F.M. (1983). Efficacy of deuterium-oxide to estimate body-composition of growing swine. *Journal of Animal Science*, **57**, 66–73.

Shields, R.G., Mahan, D.C. & Byers, F.M. (1984). In vivo body-composition estimation in nongravid and reproducing 1st-litter sows with deuterium oxide. *Journal of Animal Science*, **59**, 1239–46.

Siri, W.F. & Evers, J. (1962). Tritium exchange in biological systems. In *Tritium in the Physical and Biological Sciences*, Vienna Symp. IAEA, Vol. 2, pp. 71–80.

Smith, B.S.W & Sykes, A.R. (1974). The effects of route of dosing and method of estimation of tritiated water space on the determination of total body water and the prediction of body fat in sheep. *Journal of Agricultural Science*, **82**, 105–12.

Soberman, R., Brodie, B.B., Levy, B.B. *et al.* (1949). The use of antipyrine in the

measurement of total body water in man. *Journal of Biological Chemistry*, **179**, 31–42.
Speakman, J.R. (1997). *Doubly Labelled Water: Theory and Practice*. London: Chapman & Hall.
Speakman, J.R., Nagy, K.A., Masman, D. *et al.* (1990). Interlaboratory comparison of different analytical techniques for the determination of oxygen-18 abundance. *Analytical Chemistry*, **62**, 703–8.
Speakman, J.R., Racey, P.A., Catto, C.M.C, Webb, P.I., Swift, S.M. & Burnett, A.M. (1991). Minimum summer population and densities of bats in N.E. Scotland near the northen borders of their distributions. *Journal of Zoology (London)* **225**, 327–45.
Stanstell, M.J. & Mojica, L. (1968). Determination of body water content using trace levels of deuterium oxide and infrared spectrophotometry. *Clinical Chemistry*, **14**, 1112–14.
Tatner, P. (1990). Deuterium and oxygen-18 abundance in birds: implications for DLW energetics studies. *American Journal of Physiology*, **258**, R804–12.
Taylor, H.P. Jr (1974). The application of oxygen and hydrogen isotope studies to problems of hydrothermal alteration and ore deposition. *Economic Geology*, **69**, 843–83.
Thomas, D.L, Martin, K. & Lapierre, H. (1994). Doubly labelled water measures of field metabolic rate in white-tailed ptarmigan: variation in background isotope abundances and effect on CO_2 production estimates. *Canadian Journal of Zoology*, **72**, 1967–72.
Tiebout, H.M. III & Nagy, K.A. (1991). Validation of the doubly labelled water method ($^3HH^{18}O$) for measuring water flux and CO_2 production in the tropical hummingbird *Amazilia saucerottei*. *Physiological Zoology*, **64**, 362–74.
Torbit, S.C., Carpenter, L.J., Alldredge, A.W. & Swift, D.M. (1985). Mule deer body composition – a comparison of methods. *Journal of Wildlife Management*, **49**, 86–91.
Tsivipat, A., Vibulsreth, S., Sheng, H.P. & Huggins, R.A. (1974). Total body water measured by desiccation and by tritiated water in adult rats. *Journal of Applied Physiology*, **37**, 699–701.
Ussing, H.H. (1938). The exchange of H and D atoms between water and protein *in vivo* and *in vitro*. *Scandinavian Archives of Physiology*, **78**, 225–41.
Viljoen, J., Coetzee, S.E. & Meissner, H.H. (1988). The *in vivo* prediction of body composition in Boer goats does by means of the tritiated water space technique. *South African Journal of Animal Science*, **18**, 63–7.
Visser, G.H., Dekinga, A., Achterkamp, B. & Piersma, T. (2000). Ingested water equilibrates isotopically with the body water pool of a shorebird with unrivaled water fluxes. *American Journal of Physiology: Regulatory Integrative and Comparative Physiology*, **279**, R1795–804.
Ward, S., Scantlebury, M., Król, E., Thomson, P.J., Sparling, C. & Speakman, J.R. (2000). Preparation of hydrogen from water by reduction with lithium aluminium hydride for the analysis of delta H-2 by isotope ratio mass spectrometry. *Rapid Communications in Mass Spectrometry*, **14**(6), 450–3.
Westerterp, K.R., Wouters, L. & van Marken Lichtenbelt, W.D. (1995). The Maastricht protocol for the measurement with labeled water. *Obesity Research*, **3**, 49–57.
Williams, J.B. (1985). Validation of the doubly labeled water technique for measuring

energy metabolism in starlings and sparrows. *Comparative Biochemistry and Physiology*, **80A**, 349–53.

Williams, J.B. & Nagy, K.A. (1984). Daily energy expenditure of savannah sparrows: comparison of time-energy budget and doubly-labelled water estimates. *Auk*, **101**, 221–9.

Wolf, T.J., Ellington, C.P., Davies, S. & Feltham, M.J. (1966). Validation of the doubly-labelled water technique for bumble bees, *Bombus terrestris* (L.). *The Journal of Experimental Biology*, **199**, 959–72.

Wong, W.W. & Klein, P.D. (1987). A review of the techniques for the preparation of biological samples for mass-spectrometric measurements of hydrogen-2/hydrogen-1 and oxygen -18/oxygen-16 isotope ratios. *Mass Spectrometry Reviews*, **5**, 313–42.

Wong, W.W., Cochran, W.J., Klish, W.J. *et al.* (1988). *In vivo* fractionation factors and the measurement of deuterium and ^{18}Oxygen dilution spaces from plasma, urine, saliva, respiratory water vapour and carbon dioxide. *American Journal of Clinical Nutrition*, **47**, 1–6.

Wood, J.B., Nagy, K.A., MacDonald, N.S. *et al.* (1975). Determination of oxygen-18 in water contained in biological samples by charged particle activation. *Analytical Chemistry*, **47**, 646–50.

BRIAN T. HENEN

4

Gas dilution methods: elimination and absorption of lipid-soluble gases

Introduction

Technological advances have greatly enhanced the accuracy, speed, and portability of non-destructive body composition methods (Chapters 3–8), enabling insights into the physiology, ecology, evolution and reproductive biology of animals. Most non-destructive methods use a two-compartment model of body composition, providing direct estimates of non-lipid (or 'lipid-free') mass and estimating lipid mass indirectly, i.e. by difference: lipid mass = body mass − non-lipid mass. Since non-lipid mass frequently comprises most of the body, small errors in estimating non-lipid mass, i.e. errors as a percentage of non-lipid mass, will result in larger percent errors in lipid masses estimated by difference (Garn, 1963; Henen, 1991). Depending upon the needs of the study, non-destructive methods may provide sufficient accuracy in non-lipid and lipid mass estimates. However, some non-destructive methods may not be accurate enough for detailed analyses of energy (or other nutrient) budgets, especially for animals with small or slowly changing lipid reserves, e.g. many poikilothermic vertebrates; Henen, 1991, 1997.

Gas dilution methods may provide the requisite sensitivity and accuracy for detailed nutrient budget analyses (Lesser *et al.*, 1952, 1960; Hytten *et al.*, 1966; Henen, 1991, 1997; Gessaman *et al.*, 1998). Gas dilution methods rely primarily upon the physical properties of lipids to determine body lipid content, reducing the error introduced by first estimating non-lipid mass and then calculating lipid mass by difference. When lipid mass (M_l) is smaller than non-lipid mass (M_{nl}; $M_l < M_{nl}$), the accurate M_l estimates from gas dilution methods allow accurate calculations of M_{nl} by difference.

Measuring body composition with lipid soluble gases uses in essence, the dilution principle (Chapter 3). The primary variable to quantify is the amount of lipid soluble gas (= marker gas) dissolved in an animal's lipids, e.g. n_l, in moles. Using the solubility of that gas in lipid (α_l, moles per g lipid), M_l is calculated from $M_l = n_l \div \alpha_l$. In practice, however, a correction must be made for the amount of gas dissolved in the animal's non-lipid matter (n_{nl}, moles). This correction can be relatively minor when using gases with high lipid solubilities, e.g. cyclopropane or halothane, Tables 4.1 and 4.2. Lipid soluble gases tend to be non-polar and thus have low solubilities in polar substances such as water, body fluids, and other body components, e.g. proteins and mineral matrices. The relative solubility of a solute, e.g. gas, amongst two solvents is considered the solute's partition coefficient (PC). The ratio of lipid solubility to water solubility, the lipid-to-water partition coefficient (Table 4.1), is high for lipid soluble gases. Consequently, lipid soluble gases exposed to a body with lipid and non-lipid components will tend to dissolve into the lipids and less will dissolve into the non-lipid body components. This is why the correction for non-lipid absorption of lipid soluble gases can be minor.

The magnitude of the correction depends upon the gas partition coefficient (PC) and the animal's lipid content. An animal equilibrated with ether (lipid–blood PC = 3.32, Table 4.1), cyclopropane (lipid–non-lipid PC = 34.1) or halothane (lipid–blood PC = 88) will have roughly half of the respective gas dissolved in its lipids, and half dissolved in its non-lipid components, if the animal has lipid contents of 23.1, 2.85 and 1.1% of body mass, respectively. For example, the equation stating that the moles of cyclopropane in lipids (n_l; in moles) equals the moles in the non-lipids (n_{nl}; moles) $n_l = n_{nl}$, can be expressed, with the respective solubility coefficients and masses, as $\alpha_l \times M_l = \alpha_{nl} \times M_{nl}$. This can be arranged to the format of the partition coefficient ($PC = \alpha_l \div \alpha_{nl} = M_{nl} \div M_l = 34.1$ for cyclopropane) and used to solve the equation for the lipid content of a 100 g animal, i.e. $M_l + M_{nl} = 100$ g; $M_{nl} \div M_l = 34.1$, which would be 2.85 g or 2.85% of body mass. Thus trials using gases that are highly lipid soluble, animals with high lipid contents, e.g. >10% of body mass, or a combination thereof, will have most of the gas dissolved in the animals' lipids, the correction for n_{nl} will be small (Table 4.2, eqns. 6 and 7), and consequent errors in estimating n_l and M_l will be small (Table 4.3).

Solubility and partition coefficients have been measured in different disciplines and thus empirical data have different units (Table 4.1). However, these selected values illustrate the range and temperature

Table 4.1. *Solubility coefficients and partition coefficients (PC) for select solutes (gases) in water, lipids and other solvents*

		Solubility coefficients			Partition coefficients					
			Bunsen							
Gas	Solvent	Ostwald (mol/mol)	(ml/ml)	(ml/g)	Lipid–water	Lipid–blood	Lipid–M_{nl}	Temp. (°C)	Ref.	Comments
Helium	Water		0.00941					22.3	1	
	Olive oil		0.01370		1.46			20.5	1	a
	Water			0.0087				38.0	2	b
	Olive oil			0.0148	1.70			38.0	2	b,c
Nitrous oxide	Water	0.44						37.5	3	
	Blood	0.47				3.0		37.5	3	
	Olive oil	1.4			3.2			37.5	3	c,d
Ether	Water	15.46						37.5	3	
	Blood	15.08				3.32		37.5	3	
	Olive oil	50.0			3.23			37.5	3	d
Air	Water		0.01863					21.2	1	
	Olive oil		0.0759		4.07			19.3	1	a
Oxygen	Water	0.0272						37.0	4	
	Blood	0.0261				5.10		37.0	4	
	Olive oil	0.133			4.89			37.0	4	
	Water			0.023				38.0	2	b
	Olive oil			0.112	4.9			38.0	2	b,c
Nitrogen	Water		0.01529					22.7	1	
	Olive oil		0.0645		4.22			21.6	1	a
	Water			0.0127				38.0	2	b
	Olive oil			0.0667	5.25			38.0	2	b,c

Table 4.1 (*cont.*)

Gas	Solvent	Solubility coefficients			Partition coefficients			Temp. (°C)	Ref.	Comments
		Ostwald (mol/mol)	Bunsen (ml/ml)	Bunsen (ml/g)	Lipid–water	Lipid–blood	Lipid–M_{nl}			
Argon	Water	0.0366						20.0	4	
	n-octane	0.361			9.86			20.0	4	
	Water			0.0262				38.0	2	b
	Olive oil			0.1395	5.32			38.0	2	b
Krypton	Water	0.0522						37.0	5	
	protein	0.020						37.0	5	
	Human fat	0.425			8.14			37.0	5	c
Ethylene	Water	0.09						37.5	3	
	Blood	0.14				9.3		37.5	3	
	Olive oil	1.3			14.4			37.5	3	c,d
Xenon	Water	0.118						20.0	4	
	n-octane	4.38			37.1			20.0	4	
Cyclopropane	Water		0.203					35.0	6	
	M_{nl}			0.33				36.0	7	e
	Olive oil		9.92	11.00				37.5	8	
	Olive oil		11.00	12.18				34.0	8	
	Olive oil		13.02	14.36				27.5	8	
	Olive oil		13.95	15.35				25.0	8	
	Olive oil		10.38	11.50			34.8	36.0	8	c,f
	Olive oil		10.69	11.84	52.7			35.0	8	c,f

	Rat fat		9.78	10.84				37.5	8	
	Rat fat		10.68	11.81				34.0	8	
	Rat fat		12.67	13.95				27.5	8	
	Rat fat		10.17	11.26			34.1	36.0	8	c,f
	Rat fat		10.46	11.57	51.5			35.0	8	c,f
	Water	0.296						20.0	3	
	Water	0.278						25.0	3	
	Water	0.204						37.5	3	
	Olive oil	6.990	6.140		34.3			37.5	3	
	Cod liver oil	7.050	6.190		34.6			37.5	3	
	Paraffin oil	7.130	6.270		35.0			37.5	3	
Teflurane	Water	0.32						37.0	4	
C_2HBrF_4	Blood	0.60				48		37.0	4	
	Olive oil	29			91			37.0	4	c
Halothane	Water	0.81						37.0	4	
$C_2HBrClF_3$	Blood	2.5				88		37.0	4	
	Olive oil	220			272			37.0	4	c

Notes:
Solubility coefficients include Ostwald solubility coefficients (moles of solute per litre solvent/moles of solute per litre of gas), Bunsen solubility coefficients (ml of solute per ml of solvent; volumes at STP), and mass specific solubility coefficients (ml of solute per gram of solvent; volumes at STP). A solute's PC, the relative solubility of the solute amongst two solvents, was calculated as the ratio of the solute's solubilities in each of the two solvents, e.g. PC = lipid solubility ÷ water solubility.
a – partition coefficient estimated using solubilities at different temperatures; b – solubility units not provided in source reference; c – partition coefficient used in Figures 2 and 3; d – olive oil is the oil implied by Orcutt and Seevers 1937; e – M_{nl} equals body non-lipid mass/matter which Lesser et al. 1960 called fat-free lean mass; f–solubility estimated from regression of data in reference 8; PC from estimated α_l and the respective α_{nl} in references 6 and 7.
Source: 1. Van Slyke (1939); 2. Behnke (1945); 3. Orcutt and Seevers (1937); 4. Pollack (1991); 5. Hytten et al. (1966); 6. Lesser et al. (1952); 7. Lesser et al. (1960); 8. Blumberg et al. (1952).

Table 4.2. *Equations for calculating lipid content using the gas dilution method*

No.	Equation	
1	n_{inj}	$=P_{\mathrm{is}}V_{\mathrm{is}}R^{-1}T_{\mathrm{is}}^{-1}$
2	$n_{\mathrm{na}i}$	$=P_{\mathrm{ch}i}V_{\mathrm{eff}}R^{-1}T_{\mathrm{ch}i}^{-1}\chi_{\mathrm{cp}i}$
3	$n_{\mathrm{a}i}$	$=n_{\mathrm{re}i}-n_{\mathrm{na}i}$ [a]
4	$\alpha_{\mathrm{l}i}$	$=4.81\times10^{-3}-(1.41\times10^{-5})\mathrm{T}_{\mathrm{b}i}$ [b]
5	$\alpha_{\mathrm{nl}i}$	$=\alpha_{\mathrm{l}i}/34.1$
6	$n_{\mathrm{nl}i}$	$=W_{\mathrm{b}}\alpha_{\mathrm{nl}i}\chi_{\mathrm{cp}i}$
7	$n_{\mathrm{l}i}$	$=n_{\mathrm{a}i}-n_{\mathrm{nl}i}$
8	$M_{\mathrm{l}i}$	$=n_{\mathrm{l}i}\alpha_{\mathrm{l}i}^{-1}\chi_{\mathrm{cp}i}^{-1}$
9	$n_{\mathrm{re}(i+1)}$	$=n_{\mathrm{re}i}-(P_{\mathrm{ss}i}V_{\mathrm{ss}i}R^{-1}T_{\mathrm{ss}i}^{-1}\chi_{\mathrm{cp}i})$

Notes:
The italicized subscript i in eqns. 2 to 9 refers to the number of the gas sample removed from the chamber.
n represents the number of moles injected ($_{\mathrm{inj}}$) into the chamber, absorbed ($_{\mathrm{a}}$) or not absorbed ($_{\mathrm{na}}$) by the animal, remaining ($_{\mathrm{re}}$) in the chamber, and dissolved in non-lipid ($_{\mathrm{nl}}$) or lipid ($_{\mathrm{l}}$) body components.
P, V and T equal the pressures (atmospheres), volumes (litres) and temperatures (K) of the injection syringe ($_{\mathrm{is}}$), chamber ($_{\mathrm{ch}}$), and sample syringe ($_{\mathrm{ss}}$) at sample $_i$. V_{eff} is the effective volume (litres, see text), $T_{\mathrm{b}i}$ is the animal's body temperature (cloacal or rectal; Kelvin) and R is the ideal gas constant (0.0821 litre atm/mole K). α is the solubility coefficient (moles per gram) for lipid ($_{\mathrm{l}}$) and non-lipid ($_{\mathrm{nl}}$) body components, W_{b} is the animal's body water content (g) and $\chi_{\mathrm{cp}i}$ is the concentration (mole fraction) of cyclopropane (CP or $_{\mathrm{cp}}$) inside the chamber at sample $_i$. $M_{\mathrm{l}i}$ is the estimate of lipid content (g) at sample $_i$.
[a] For the first sample, the moles remaining in the chamber (n_{re_1}) equals n_{inj}
[b] regression calculated from Blumberg et al. (1952).
Source: From the validation using cyclopropane, Henen (1991).

dependence of coefficients for several solutes and solvents. They are also critical for evaluating the sensitivity of gas dilution methods to analytical errors and assumptions (see Application and sensitivity analysis – gases, and Validations of the gas dilution method – solubility coefficients). Using molar solubility coefficients, e.g. moles of solute per g of solvent (Table 4.2), allows lipid content to be estimated in common units, i.e. g of lipid, and solute quantity to be expressed in common units (moles) easily converted to other units (ml) and conditions, e.g. STP.

The use of lipid soluble gases to estimate body composition non-destructively has been applied through measuring the elimination of lipid-soluble gas from the body. Behnke (1945) summarized studies on humans and dogs where the total volume of N_2 gas, e.g. 1076 ml, eliminated from the lungs in 12 h closely approximated the total volume, e.g. 1145 ml, calculated from (i) the 'fat' and 'fluid' solubility coefficients for N_2 and (ii) the subjects' 'fat' and 'fluid' masses, respectively. This suggested that body composition could be determined by quantifying the amount of body N_2 eliminated through the lungs into respirometers. However, Behnke (1945) also provided evidence that cutaneous absorption or elimination, especially through cutaneous lesions, significantly altered the volume of respiratory N_2 eliminated and altered the estimates of body composition. If N_2 had a higher lipid–water PC, the body composition estimates may have been more accurate. Quantifying the elimination of less common gases (relative to atmospheric N_2 levels) with lipid–water PC higher than for N_2 may provide accurate, non-destructive measures of body composition.

More recently, the absorption of lipid soluble gases with higher PC has been used to accurately estimate the body composition of a variety of animals (Lesser *et al.*, 1952, 1960; Hytten *et al.*, 1966; Henen, 1991, 1997; Gessaman *et al.*, 1998). To quantify the absorption of the marker gas, a subject is fitted with a special breathing apparatus (see Hytten *et al.*, 1966 and references therein) or placed inside an equilibration chamber, e.g. Fig. 4.1. A known amount of marker gas (n_{inj}, eqn. 1, Table 4.2) is then introduced into the system (chamber or breathing apparatus, plus the animal) and allowed to diffuse throughout the system until the gas has reached an equilibrium state in the animal and its surrounding space within the system. Some of the marker gas remains inside the system but outside of the animal, so a correction must be applied for the amount of administered gas that was not absorbed by the animal, e.g. moles not absorbed, n_{na}, eqn. 2, Table 4.2. After this correction, the absorption method resembles the elimination method for lipid soluble gases. The moles of gas absorbed by the body (n_a, eqn. 3, Table 4.2) is corrected for n_{nl} (eqns. 4 to 7, Table 4.2), resulting in the quantity of gas absorbed by the lipids (n_l). Then n_l and α_l are used to calculate M_l (eqn. 8). The sampling and calculations are repeated, correcting for the amount of marker gas removed by sampling (eqn. 9, where i represents the number of the gas sample removed from the chamber), until the body composition estimates are within a certain level of precision, e.g. a coefficient of variation $<10\%$; Henen, 1991.

Despite the additional correction for n_{na}, validations demonstrate that the gas dilution method provides extremely accurate estimates of M_l and M_{nl} (see below). This accuracy was probably due, in part, to the high lipid solubilities and lipid-to-water partition coefficients of the gases used as

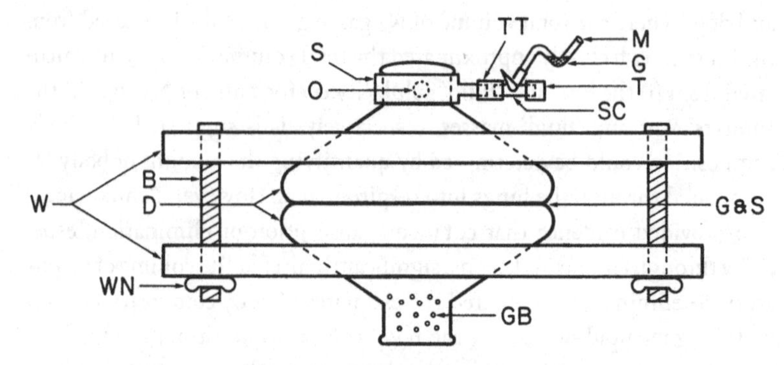

Fig. 4.1. Equilibration chamber for quantifying body lipid mass by gas dilution methods. Chamber consists of two glass desiccator lids (D) held together by a plywood ring clamp (W), wing nuts (WN) and bolts (B). Desiccator sleeve (S) and three-way metal stopcock (SC) can align with chamber opening (O) for injecting gas into chamber and sampling chamber gas through the stopcock and Tygon tubing (T). Teflon tubing (TT) connects the stopcock to the sleeve. Chamber pressure is monitored through a glass manometer (M) containing glycerol (G). Glycerol is used to seal the chamber and sleeve connection and a glycerol and starch (G and S) mixture helps seal the connection of the chamber halves. Inert glass beads (GB) can fill space to minimize headspace or chamber size relative to animal size. Water vapour and carbon dioxide absorbents can also be placed inside the chamber. (Figure from Henen, 1991.)

well as efforts to minimize the headspace of the equilibration systems. The gas dilution method is not as quick and simple as are other non-destructive methods, e.g. bio-impedance and total body electrical conductivity. However, the detail and complexity of the gas dilution method typically makes gas dilutions more accurate, and less sensitive to animal shape, hydration state and body temperature (Henen, 1991) than other non-destructive methods. The following review of the application, sensitivity analyses, and validations of the gas dilution method reveals how the method can be extremely accurate and less sensitive to nuisance variables than other non-destructive methods.

Application and sensitivity analyses

Gas dilution methods require excellent analytical skills to apply effectively. Developing good analytical technique and dexterity in model situations, i.e. practising the method upon inanimate models such as containers of olive oil, facilitates applying the method to live animals. The following discussion concerns implementing the gas dilution method through monitoring the absorption of lipid soluble gases. This discussion combines knowledge from (i) completing sensitivity analyses to evaluate the potential errors of the gas dilution method (Henen, 1991; Table 4.3, Figs. 4.2 and 4.3), (ii) validating the use of cyclopropane gas to determine body composition (Henen, 1991; Table 4.4), (iii) applying the method to wild animals (Henen, 1997), and (iv) other successful validations (Lesser *et al.*, 1952; Gessaman *et al.*, 1998). Some of the major concerns regarding the gas dilution method include: (i) choice of lipid soluble gas, (ii) integrity and size of the equilibration chamber, (iii) ease of use and stability of chambers, (iv) gas injections and sampling, and (v) gas analysis.

Gases

Which gas is used is one of the most important considerations in applying the gas dilution method. The selected gas affects the accuracy of the body composition estimates, whether new solubility coefficients must be determined, how the gas concentration will be analysed and which gas concentration should be used. It may also affect whether the body may metabolize the gas, whether radioisotope permits are required, and whether fire or health precautions must be considered. The primary advantage of using lipid soluble gases to measure animal lipid content is that most of the gas is diluted into the animals' lipids. The quantity of gas absorbed by non-lipid components (n_{nl}) reduces the quantity of gas dissolved in the animals' lipids (n_l), so a correction is necessary. Analogous to the problem of calculating lipid mass by difference when $M_l < M_{nl}$, the error in n_{nl} and n_{na}, especially if large, causes errors in the calculated n_l. Gases with high partition coefficients (PC) should result in most of the gas dissolving into the lipids such that errors in n_{nl} and n_{na} have a small influence on the accuracy of n_l. This was demonstrated with (i) a computer model of the calculations for cyclopropane gas absorption (Henen, 1991), and (ii) a similar model evaluating lipid estimates when using gases with different partition coefficients (Figs. 4.2 and 4.3). Dilutions using gases with high PC, e.g. cyclopropane (Table 4.1), are less sensitive to measurement errors than dilutions using gases with low PC, e.g. ether, because dilutions with high PC gases result in small n_{na}, small n_{nl} and large n_l. In addition, the model illustrates that there is a decreasing return in choosing gases with higher lipid–water PC (Fig. 4.3). These models, however, do not indicate whether equilibration times differ amongst gases.

The solubility coefficients and lipid–water partition coefficients have been measured for a few gases (Table 4.1), so using another gas necessitates

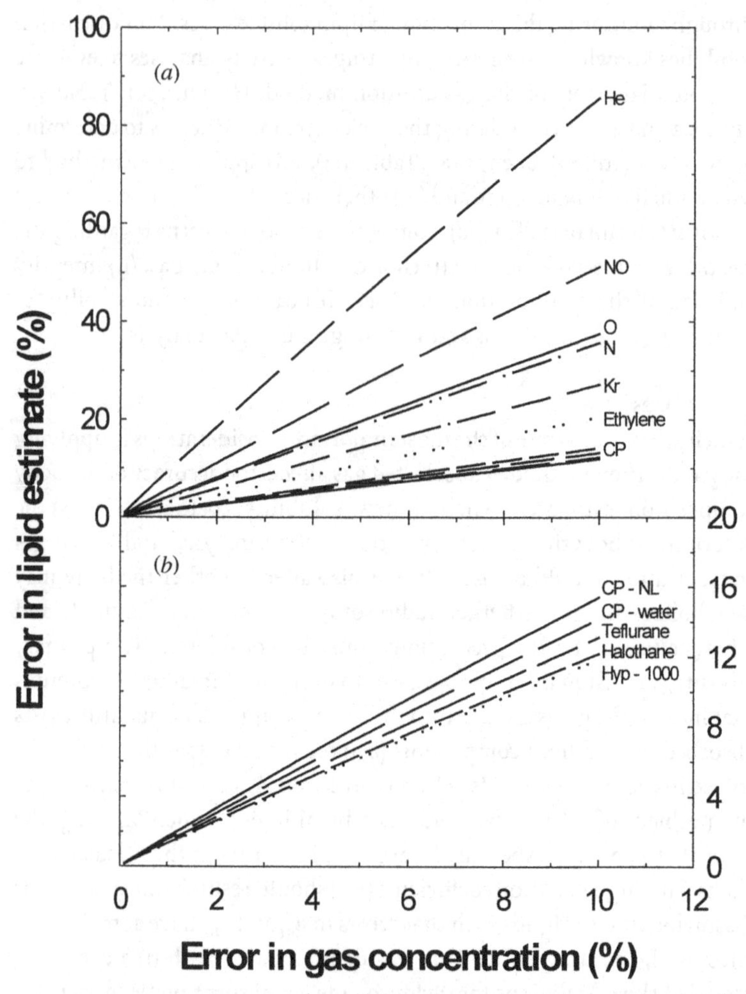

Fig. 4.2. Computer model of errors in lipid estimates resulting from errors in measuring equilibrium concentrations of lipid soluble gases. (*b*) Shows an expanded view of the results for gases with partition coefficients (PC) equal to, or greater than, those for cyclopropane gas. The model simulated a 100 g organism containing 5 g of lipid and utilized gases having different lipid-to-water PC (see Table 4.1). Errors in lipid estimates were modelled using lipid-to-water PC for helium (He), nitrous oxide (NO), oxygen (O), nitrogen (N), krypton (Kr), ethylene, cyclopropane (CP), teflurane, halothane and hypothetical gases having PC of 10^3 (Hyp -1000), 10^{12} and 10^{99}. The results for hypothetical gases with PC $= 10^{12}$ and 10^{99} were indistinguishable, and thus not illustrated, from results for Hyp -1000. The PC for cyclopropane included those for olive oil-to-water (CP in (*a*)), rat fat-to-non-lipid matter (CP – M_{nl}, part b) and rat fat-to-water (CP – water in (*b*)). For any given measurement error in gas concentration, the errors in lipid estimates differed negligibly between the cyclopropane models for olive oil-to-water PC and rat fat-to-water PC.

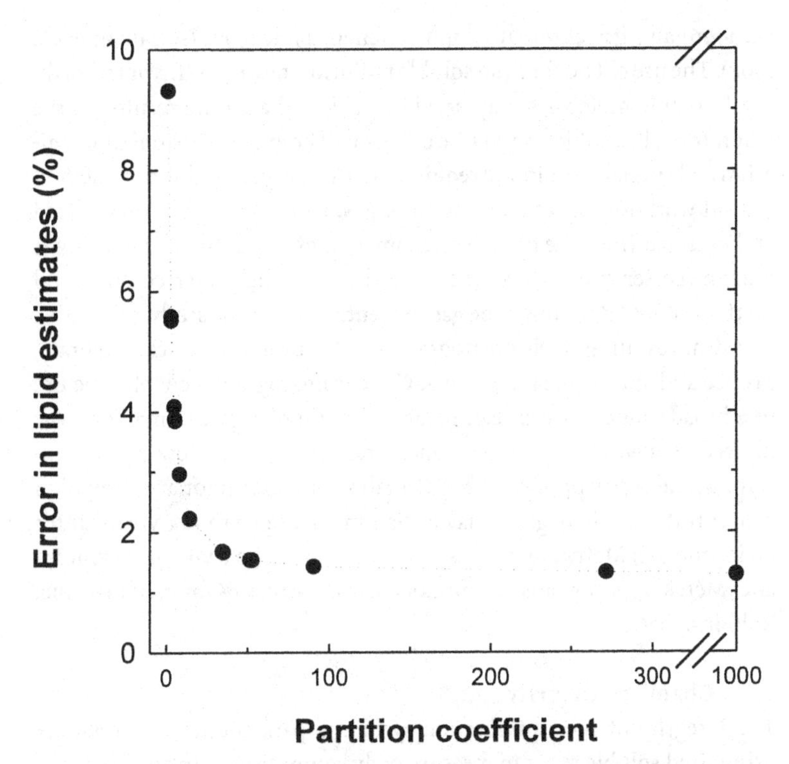

Fig. 4.3. Relationship between errors in lipid estimates and the lipid-to-water partition coefficients of gases illustrated in Fig. 4.2. The errors in lipid estimates were those that the model provided for the 1% errors in gas concentrations.

determining (i) the lipid and non-lipid solubility coefficients for that gas, and (ii) the relationships of gas solubilities to temperature (Table 4.1). The solubilities of cyclopropane for olive oil, rat fat and other oils are similar, suggesting that substituting the cyclopropane coefficient of one lipid, e.g. rat fat, for that of another lipid introduces only small errors. This has been confirmed in validation studies where the cyclopropane solubilities for rat fat and human 'fat-free lean mass' (Blumberg *et al.*, 1952; Table 4.1) were used to accurately estimate the body composition of turtles (Henen, 1991) and pigeons (Gessaman *et al.*, 1998).

Analytical method for gases

The analytical methods for quantifying gas concentrations can range from complex combustion and titration methods (Lesser *et al.*, 1952) to simpler methods including liquid scintillation counting (LSC; Hytten *et al.*, 1966) and gas chromatography (Henen, 1991, 1997; Gessaman *et al.*, 1998). The use of LSC for lipid soluble radioisotopes, e.g. ^{85}Kr, or radiolabelled organic molecules may provide quick and accurate results (in the laboratory), but subjects and researchers will be exposed to ionizing radiation and special permits are required. In theory, the gas dilution method should work for any concentration of gas. Unless the gas is catabolized and excreted from the body, a phenomenon of which little or nothing is known (Lesser *et al.*, 1960), the lower limit of the preferred analytical method should determine the gas concentration for the study.

Advances in gas chromatography (GC) promote quick, accurate, precise and simple measurements. Gas chromatography enables the use of a broad range of gases, e.g. polar and non-polar gases; small or large molecules, a wide range of gas concentrations, and very low gas concentrations, i.e. a few ppm or ppb. GC analyses are useful for avoiding high concentrations where gases may be flammable, e.g. 2 to 10% v/v in air for cyclopropane; Merck, 1952, cause anesthesia (15 to 30% v/v for cyclopropane; Merck, 1952) or cause respiratory arrest (35 to 40% for cyclopropane; Robbins, 1936).

Chamber integrity

The integrity of the equilibration system is paramount to successfully using lipid soluble gases to measure body composition. Errors due to gas uptake or loss from the equilibration system can plague a gas dilution study. When gas elimination is monitored (Behnke, 1945) or breathing apparatuses are used for gas absorption (Hytten *et al.*, 1966), cutaneous exchange of the marker gas can introduce large errors (but see Lesser *et al.*, 1960). When equilibration chambers are used for monitoring gas absorption, lipid soluble gas can leak from the chamber, causing misleadingly low gas concentration measurements and high lipid estimates (Table 4.3). If the atmosphere is not cleared of the marker gas at the beginning of the trial, excess marker gas will reside in the chamber at equilibrium, resulting in misleadingly high concentrations and low lipid estimates.

Variations of the glass equilibration chamber used upon laboratory rats (Lesser *et al.*, 1952) were used successfully upon turtles (Henen, 1991), tortoises (Henen, 1997) and pigeons (Gessaman *et al.*, 1998), indicating that this design resists leaks while allowing quantitative injections into and sampling from the chamber. Glass, and perhaps metal, may be the best material for the equilibration chambers. Lipid soluble gases readily dissolve into and through rubber, plastics, oil, grease and other non-polar

Table 4.3. *Sensitivity of gas dilution method to 1% overestimates (due to analytical error) in measurement variables (symbols indicated in Table 4.2) for a model cyclopropane dilution*

Measurement variable	Error in lipid estimate (%)
P_{ch}, χ_{cp}	−4.3
P_{is}, V_{is}	4.3
V_{eff}	−2.9
V_{an}	1.2
α_l	−1.0
T_b	0.89
α_{nl}, W_b	−0.40
T_{is}	−0.36
T_{ch}	0.26
P_{ss}, V_{ss}	−0.0034
T_{ss}	0.0003

Notes:
Positive and negative errors indicate that M_l was overestimated and underestimated, respectively. Temperatures were overestimated on the scale of measurement (degrees Celsius, e.g. 0.25 °C for a 25.0 °C measurement) and then converted to Kelvin for the sensitivity analysis. If W_b was calculated from body mass and per cent body water, the same error would result from 1% overestimates of body mass or per cent body water.
Source: Data from Henen (1991).

substances (see Lesser *et al.*, 1952), but not glass, metal and Teflon. Glass chambers also allow visual monitoring of subjects during the equilibration. Additionally, the chambers must have one or more ports for (i) quick and quantitative injection of the marker gas into the chamber, (ii) quick and representative sampling of chamber gas for analysis of gas concentration, (iii) maintaining oxygen levels within the chamber, and (iv) maintaining chamber pressure at atmospheric levels. These ports, plus the seal of the chamber lid or halves (Fig. 4.1), may leak the lipid soluble gas. A mixture of glycerol and starch forms an adequate seal at the chamber halves and seals connections between the sleeve, injection-sampling port and manometer (Fig. 4.1). Glycerol lubricates the sleeve and stopcock valve, and serves well as the fluid in the manometer that is used to monitor and help maintain chamber pressure (Henen, 1991). Teflon tubing, with Teflon tape wrapped on the sleeve–stopcock joint to ensure a tight physical seal (Fig. 4.1), also serves as a connecting material that resists leaks. The sleeve design allows the chamber to be sealed from the injection-sampling port and manometer during most of the equilibration, helping to minimize leaks. Oxygen consumption by the animal, plus water and carbon dioxide absorption by anhydrous $CaSO_4$ and NaOH pellets, respectively, tends to reduce chamber pressure below atmospheric pressure. This facilitates the addition of oxygen to the chamber through the manometer or injection port, plus the 'negative pressure' helps tighten the seals and reduce leakage.

Chamber size

To help minimize errors introduced when correcting for n_{na}, it is important to increase the proportion of marker gas absorbed by the animal. This is facilitated by minimizing the headspace or chamber size relative to animal size, which can be accomplished by (i) selecting chambers that approximate the size and shape of the organism, e.g. cylindrical for rats and pigeons, Lesser *et al.*, 1952 and Gessaman *et al.*, 1998; and elliptical for turtles, Henen, 1991, and (ii) adding inert objects like glass beads to fill excess chamber space (Fig. 4.1). However, means of mixing chamber gases (see Gas sample concentration) must be considered when choosing chambers and inert, space-filling objects. Chamber volume (V_{ch}) can be measured gravimetrically; the mass of water required to fill the chamber is converted to water volume using the water density at the water temperature (Henen, 1991; Gessaman *et al.*, 1998). V_{ch} can also be calculated from diluting a known volume (V_{inj}) of pure marker gas (mole fraction concentration, $\chi = 1$) in the chamber, and measuring the gas concentration at equilibrium ($V_{ch} = \chi \times V_{inj} \div \chi_{eq}$; Gessaman *et al.*, 1998).

When calculating n_{na}, it is necessary to know the volume of air space inside the chamber that excludes the lipid and non-lipid components of the animal. This effective volume (V_{eff}, eqn. 2, Table 4.2) equals the chamber volume minus the volume of the animal (V_{an}) and other chamber contents, e.g. glass beads, magnetic stir bar, $CaSO_4$ plus its inert screen wire container, and NaOH plus its inert Teflon container. V_{eff} includes the volume of air in the animal's respiratory and gastrointestinal (GI) systems so estimates of V_{an} must exclude these respiratory and GI volumes. The errors in V_{eff} and V_{an} may partially cancel each other when calculating M_l (Table 4.3). Respiratory gas volumes, but not GI gas, can be estimated by quick dilutions of another inert gas, e.g. helium, Hytten *et al.*, 1966. Alternatively, V_{an} can be calculated from body mass and body density, i.e. $V_{an} = M_b/\rho_b$, where M_b is body mass in g and ρ_b is body density in g per ml;

Table 4.4. *Accuracy of estimating lipid mass (M_l) and nonlipid mass (M_{nl}) from the absorption of cyclopropane gas, body water dilution, and total body electrical conductivity (TOBEC)*

	Lipid mass		Non-lipid mass		
Organism	*r*	Error, %	*r*	Error, %	Reference
Cyclopropane					
Turtles	0.989	10	1	0.18	Henen (1991)
Pigeons		11			Gessaman *et al.* (1998)
Rats	0.989	5.6	0.998	0.78	Lesser *et al.* (1952)
Total body water dilution					
Turtles[a]	0.658	280	0.946	4.2	Henen (1991)
Total body electrical conductivity (TOBEC)					
Rabbits	0.88	26	0.989	3.1	Klish *et al.* (1984)

Notes:
Accuracy is expressed as (i) the correlation coefficient (*r*) between non-destructive estimates of M_l (and M_{nl}) and chemically extracted values, and (ii) percent error between estimates and extracted values. The error in M_{nl} was not reported in Gessaman *et al.* (1998). See Henen (1991) for original analysis of turtle, rat and rabbit data.
[a] Lipid mass and estimates by TBW dilution were not correlated ($P = 0.18$, Henen, 1991).

Henen, 1991). Henen (1991, 1997) measured V_{an} that were free of respiratory and GI gas volumes by weighing volumes of homogenized carcasses. The density of lipids (ρ_l) and a 'lipid-free body' (ρ_{nl}) in humans is 0.9 and 1.1 g/ml, respectively (Keys & Brozek, 1953; Brozek *et al.*, 1963; Forbes, 1978). For pond turtles (*Trachemys scripta*; Henen, 1991) and desert tortoises (*Gopherus agassizii*; Henen, 1997) with low lipid levels (*c.* 1 to 8% of M_b), ρ_b equalled 1.17 and 1.13 g/ml, respectively, so a theoretically 'lipid-free' turtle or tortoise should have a density of roughly 1.2 g/ml. The potential range of ρ_b (0.9 to 1.2 g/ml) is much narrower for most animals because their lipid contents are much less than 50% of M_b. Thus it is difficult to overestimate or underestimate ρ_b by more than a few percent.

Successful validations on turtles and pigeons used chamber volumes where $V_{eff} \div V_{an}$ equalled 2 to 4 and 10, respectively (Table 4.4). The large chamber volumes used for pigeons, relative to those for turtles, may have been possible due to the pigeons' high lipid content (5 to 22% of body mass) relative to that for turtles (0.71 to 3.5%). Also, the large chamber volumes for pigeons may have been necessary to supply the high oxygen volumes to support the pigeons' endothermic metabolic rates. Chamber oxygen tensions can be boosted to support aerobic metabolism without apparent complications of chamber gas analyses (Henen, 1991, 1997; Gessaman *et al.*, 1998), but some flame ionization detectors (FID) on gas chromatographs extinguish at oxygen tensions exceeding 30% (Henen, 1991, 1997). If FID flames extinguish, the gas chromatograph must be restarted and calibrated, and chamber oxygen tensions must be allowed to decline before analysing more gas samples.

Chamber use and stability

The gas dilution method requires many chamber manipulations and measurements such as maintaining and measuring chamber pressure and temperature (P_{ch} and T_{ch}, respectively). However, a well-designed chamber (Fig. 4.1) can facilitate these manipulations and measurements, enhancing the precision and accuracy of the measurements and body composition estimates (see sensitivity analysis in Table 4.3). P_{ch} and T_{ch} are recorded for every chamber gas sample and are critical to accurate estimates of M_l and M_{nl} (eqn. 2, Table 4.2).

Maintaining P_{ch} at or below atmospheric pressure helps to prevent chamber leaks. The water and carbon dioxide absorbents, plus oxygen consumed by the animal, will maintain P_{ch} below atmospheric pressure if T_{ch} does not increase significantly during the equilibration (see below). For the calculations, we assume that P_{ch} equals the atmospheric pressure when the gas samples are collected so judicious use of the chamber manometer is critical. The manometer is extremely sensitive to pressure changes and the manometric inspiration of oxygen or air into the chamber can maintain P_{ch} very close to atmospheric pressure. Administering oxygen through the injection port can help maintain P_{ch} as well as chamber oxygen tension. The three-way stopcock allows quick switching between the manometer, injection-sampling port, and the closed position as necessary for (i) assessing P_{ch}, (ii) completing injections (marker gas and oxygen) or sampling chamber gas, and (iii) testing for and reducing leakage during the equilibration, respectively. P_{ch} can be monitored without complications, e.g. leaking chamber gas through the manometer or displacing the manometer fluid by (i) closing the stopcock, (ii) quickly opening and closing the chamber – sleeve connection by twisting the sleeve, and (iii) using the manometer to monitor the pressure within the sleeve ($\simeq P_{ch}$). The high arm of the manometer prevents the drawing of glycerol into the stopcock. Since standard or electronic manometers resolve atmospheric pressure to 0.1 to 1 torr, which is typically $<0.2\%$ of atmospheric pressure (Henen, 1991), these devices should provide sufficient accuracy for P_{ch} and ultimately, body composition estimates (Table 4.3).

Accurate chamber temperatures (T_{ch}) are critical because T_{ch} is used to calculate n_{na}. For ectotherms, T_{ch} is also important in estimating T_b, and thus solubility coefficients (eqns. 4 and 5, Table 4.2) when deciding whether equilibrium has been reached. The true T_b (cloacal or rectal, measured immediately after reaching equilibrium; Henen, 1991; Gessaman *et al.*, 1998) is used when completing the final calculations. However, if T_{ch} differed slightly during the final gas samples, T_b was assumed to have changed by the same amount for turtles and tortoises (Henen, 1991, 1997). For example, if T_{ch} decreased 0.2 °C between the two final samples, then the T_b for the penultimate sample equalled the rectal–cloacal temperature plus 0.2 °C.

Gas dilution estimates of body composition are less sensitive to errors in T_{ch} than to errors in P_{ch} (Table 4.3). The sensitivity analysis summarized in Table 4.3 evaluated the effect of 1% errors from the scale of the measurement, i.e. degrees Celsius for T_{ch}, T_{is}, T_{ss}. However, these temperatures are measured accurately in Celsius and are converted to Kelvin for the calculations. Measurement errors on the Celsius scale are relatively small in Kelvin, and may explain why temperature errors have small effects on body composition estimates. Inserting the accurate thermocouple leads under foam or fiberglass insulation on the chamber (Henen, 1991, 1997), or into the chamber (Gessaman *et al.*, 1998), can help thermally stabilize the leads and T_{ch} measurements. However, unstable environmental temperatures can wreak havoc upon equilibrations. Rapid increases in T_{ch} cause P_{ch} to increase considerably, which often destroys the chamber seals and displaces the manometer fluid, resulting in leakage of the marker gas and overestimates of M_l.

When ambient temperature shifts, the standard curves of poorly insulated gas chromatographs shift and must be reanalysed. The temperature sensitivity of equilibration chambers and some chromatographs can limit the portability of the technique. Additionally, endotherms may have difficulty thermoregulating inside the equilibration chamber. In a validation study using pigeons (Gessaman *et al.*, 1998), large pigeons had elevated T_b (*ca.* 45 °C) compared with smaller pigeons ($T_b \sim 41$ °C) and some large pigeons died from hyperthermia. A temperature controlled room or trailer will minimize or eliminate many of the temperature related problems, but direct cooling of the chamber walls (Gessaman *et al.*, 1998) may be necessary to reduce hyperthermia.

Gas injections and sampling

Inaccurate and imprecise gas injections and sampling of chamber gas may cause the largest errors in M_l estimates through (i) injecting erroneous quantities of marker gas into the chamber (P_{is} and V_{is}; eqn. 1, Table 4.2) and (ii) errors in the gas concentration of samples (χ; eqns. 2, 6, and 9 in Table 4.2; see also Table 4.3). Errors in the volume of marker gas injected (V_{is}, $_{is}$ represents injection syringe) can be minimized through good analytical methods and equipment. First, gas-tight syringes, e.g. Pressure-Lok, A-2; Precision Sampling, Baton Rouge, LA, USA, have precise glass barrels, Teflon plungers, Teflon fittings for the needles, and a Teflon valve that closes the syringe. These syringes have very low leakage rates that facilitate accurate and precise injections and samples. After extensive use, however, the plunger tips wear and must be replaced to minimize leaking.

The volumes of the injection and sampling syringes (V_{is} and V_{ss}, Table 4.2) can be calibrated gravimetrically to well within 1% accuracy. Furthermore, using one well-calibrated syringe, at one well-calibrated volume, for all injections minimizes introducing syringe calibration errors. Select an appropriately sized syringe for the particular injection volume. It is difficult to accurately calibrate large syringes, e.g. 10 ml, for small volumes, e.g. 100 μl, and errors from repeatedly injecting small volumes, e.g. 100 μl, may exceed the error in the injection of one large volume, e.g. 800 μl. The same principles apply to the sample syringes although V_{ss} contributes considerably less error to body composition estimates (Table 4.3).

The pressure in the injection and sampling syringes (P_{is} and P_{ss}, respectively) can be closely approximated by atmospheric pressure. The stock of lipid soluble gas (99.0% purity) is maintained at several atmospheres of pressure in a lecture bottle or gas cylinder. A gas-tight syringe, fitted with a side-port needle that prevents needle clogging, is used to withdraw the stock gas and the gas pressure must be quantified (eqn. 1, Table 4.2). The simplest method is to bring the gas pressure in the syringe to atmospheric pressure. This is accomplished by placing the port of the needle a few mm underwater, opening the syringe valve and purging the syringe until the bubbling ceases. The valve of the gas-tight syringe is closed immediately. Atmospheric pressure and syringe temperature (T_{is}) are recorded. In reference to purging the syringe underwater, the hydrostatic pressure of 3 mm of water is very small (0.22 mm Hg), so the gas pressure within the purged syringe, i.e. P_{is} should be very close to atmospheric pressure. The gas can then be injected into the chamber and n_{inj} can be calculated using the Ideal Gas Law (eqn. 1, Table 4.2).

The pressure of the gas in the sample syringe (P_{ss}) should be extremely close to atmospheric pressure, but may be slightly lower because sampling removes some gas from the chamber. The sample volumes necessary for GC analysis, however, are typically very small (ca. 50 to 250 μl) relative to chamber volumes (*ca.* litres), so P_{ss} deviations from atmospheric pressure are usually negligible. Likewise, the removal of chamber gas by sampling may have negligible effects upon n_{re}, the moles of marker gas remaining in the chamber. Thus, correcting for marker gas removed by sampling (eqn. 9, Table 4.2) may not significantly affect body composition estimates. Alternatively, one can sample a larger volume than is necessary for GC analysis, e.g. collect 60 μl for a 50 μl analysis, press the plunger to match the proper analytical volume (increasing gas pressure in the syringe) and purge the syringe to atmospheric pressure prior to analysis.

The errors in syringe temperatures typically produce small errors in body composition estimates (Table 4.3). As for T_{ch}, this is likely due to accurate thermocouples and using the Kelvin scale in the calculations. Still, errors in syringe temperatures can be reduced simply. Handling the syringes causes temperature differences along the barrel. Minimal handling of the syringe, plus a small piece of insulating foam on the barrel, can reduce these local, transient temperature fluctuations.

A reduced P_{ch} can facilitate quantitative transfer of the lipid soluble gas into the chamber. First, a large syringe (plastic or glass) is used to withdraw air from the chamber. Then the stopcock, plus the sleeve if the chamber has this design (Fig. 4.1), is closed. This procedure can be repeated to further reduce P_{ch} prior to injecting the marker gas. Then the injection syringe is mounted on the injection port (through the tight-fitting Teflon or Tygon tubing), the stopcock, sleeve and syringe valve are opened and the syringe is emptied several times through repeated plunging without removing the syringe from the port. For a more quantitative transfer, air can be allowed in slowly through the manometer, between plunges of the injection syringe, to help clear marker gas from the sleeve for subsequent plunges.

Gas sample concentration

Collecting a sample representative of the chamber gas at equilibrium is critical in determining equilibrium gas concentrations (χ) and body composition (eqns. 2, 6, 8 and 9 of Table 4.2; and Table 4.3). Representative samples can be collected if the chamber gas is well mixed by Teflon-coated stir bars powered by external electric stirrers (B. T. Henen, personal observations; Gessaman *et al.*, 1998). The chamber gas can also be mixed by attaching a large glass syringe (gas-tight, or a simple glass syringe with glycerol lubricating the plunger) to the sample port, and pumping the plunger several times (Henen, 1991). The latter facilitates mixing within the chamber during the equilibration, and mixing within the sleeve and sampling port prior to sampling. When sampling, the gas-tight sample syringe is attached at the tight-fitting port and the plunger is pumped several times to ensure that the sample represents the chamber gas. The syringe valve is closed, T_{ss} and atmospheric pressure are recorded, and the sample is ready for GC analysis (pending purging to atmospheric pressure if necessary).

Sample analysis by gas chromatography is very simple. The sample is injected, using the gas-tight syringe and needle, through the septum of the chromatograph's injection port. The chromatograph output, e.g. peak height or area is then compared to a standard curve generated for the GC system on that day. The time required to receive the results varies with the gas, the gas chromatograph and the type of GC column. For example, analysing one cyclopropane sample required 45 s for one gas chromatograph (Henen, 1997) and 4 min for another chromatograph (Henen, 1991).

Theoretically, the concentration of lipid soluble gases should not affect the accuracy of the method unless the animals can catabolize the gas. Catabolism has not been detected for the levels of cyclopropane used successfully to date (250 000 ppm, Lesser *et al.*, 1952; 10 000 to 20 000 ppm, Lesser *et al.*, 1960 and Gessaman *et al.*, 1998; 5000 to 7000 ppm, Henen, 1991; 300 to 600 ppm, Henen, 1997), but little is known about the catabolism of cyclopropane. GC systems are accurate at low concentrations (ppm to ppb), obviating the use of higher, potentially unhealthy, concentrations. Finally, the equilibrium concentrations should match the most accurate, linear range of the gas chromatograph.

An example for tortoises illustrates how to prepare and analyse gas standards (Henen, 1997). Standards were prepared using 50-ml serum vials capped with Teflon Mininert valves (Alltech Associates). These valves resemble the Teflon valves of the gas-tight syringe, except they have a replaceable septum to minimize gas leaks while inserting gas into, or withdrawing samples from, the serum vials. The volumes of the serum vials, capped with matched Mininert valves, were calibrated gravimetrically with water. Using a calibrated, 100 μl gas-tight syringe, 25 to 55 μl of pure cyclopropane (99.0%, purged to atmospheric pressure; Scott Specialty Gases) were diluted into the standard vials. Syringe temperature was recorded when the syringe was purged to atmospheric pressure. Vial temperature and atmospheric pressure were recorded when the vial was sealed prior to the dilution. The pressures, volumes and temperatures of the syringe and vials were used to calculate the concentration (χ, mole fraction) of the standard using the Ideal Gas Law. The standards were collected and analysed in the same manner as were the chamber samples. The concentrations of the chamber samples were calculated from the one or two standard curves generated that day.

Other considerations

A few other variables must also be considered when using this method. Although the method relies primarily upon the properties of lipids, corrections for n_{nl} can be substantial. Corrections for n_{nl} that use M_{nl} estimated from body fluids or total body water (W_b, g) seem to provide satisfactory results (Behnke, 1945; Lesser *et al.*, 1952; Henen, 1991, 1997; Gessaman *et al.* 1998). However, this may be largely due to the low gas solubilities for water and non-lipid matter such that errors in W_b and α_{nl} have relatively minor effects upon n_l and the estimated M_l (Tables 4.2 and 4.3). W_b can be estimated as a percentage of body mass, from correlations to body mass, or by dilution of isotopically labelled water (Henen, 1991; Chapter 3). The isotopic water dilution can be completed prior to, during, or after the gas equilibration. W_b can be estimated to within ~2 to 6% accuracy using oxygen-18 and tritium dilution spaces, respectively (Henen, 1991). This level of accuracy should cause relatively small errors (1 to 2.5%) in lipid estimates for an animal that is 5% lipid by mass (Table 4.3) and even smaller errors for animals with higher lipid contents.

Endotherms may have difficulties thermoregulating in equilibration chambers. A combination of a breathing apparatus plus a system for quantifying cutaneous losses of marker gas may present less hazards than imposed by small equilibration chambers. Ectotherms appear to lack the thermoregulatory problem associated with equilibration chambers (Henen, 1991, 1997). The lower T_b of ectotherms may increase the accuracy of the gas dilution method for ectotherms. A lower T_b translates to higher gas solubilities (Tables 4.1 and 4.2), favouring greater absorption of marker gas by the body, i.e. smaller n_{na}. The small n_{na} at low temperatures should result in small errors in M_l estimates. This temperature effect may help explain the high accuracy obtained for the validation of cyclopropane gas dilutions upon turtles having low T_b and low lipid levels (0.7 to 3.5% of body mass). The body temperatures of turtles during equilibration (~ 23 to 31 °C) were below typical endotherm T_b.

Validations of the gas dilution method

Accuracy

The gas dilution method has been applied to tortoises (Henen, 1997), humans, and dogs (Behnke, 1945; Lesser *et al.*, 1960; Hytten *et al.*, 1966; Forbes, 1978). Still, there are only three validations of the gas absorption method with direct chemical analysis of carcasses (Lesser *et al.*, 1952; Henen, 1991; Gessaman *et al.*, 1998). All three validations used cyclopropane as the lipid soluble gas and glass equilibration chambers to complete the dilutions. The cyclopropane estimates of M_l matched the M_l determined by carcass analysis to within 11%, on average, for all three validations (Table 4.4). Furthermore, M_{nl} estimated by difference matched those values determined by carcass analysis to within 1% for turtles and rats (Table 4.4).

Accuracy can be expressed in different ways. Although high correlation coefficients (*r*; Table 4.4) are frequently used in comparing non-destructive and destructive measures of body composition, r values without the slopes and intercepts of the correlation are not very helpful. The validation upon turtles (Henen, 1991) indicated that the cyclopropane estimates and destructive measures of M_l had a high correlation coefficient, a slope not different than one, and an intercept not different from zero. Also, the slope and intercept for the turtle correlation did not differ from those values for the validation upon rats (Lesser *et al.*, 1952). A useful expression of a method's accuracy is the average error (%) of the non-destructive estimates relative to the destructive values (Table 4.4). This value is useful when deciding which method is appropriate for a particular application. That is, how much error in lipid mass (or non-lipid mass) estimates do I expect to obtain when I use this method?

Estimates of turtle body composition by body water dilution were relatively poor (Table 4.4). Water dilution estimates of M_l were not correlated to chemically extracted values and average error was very large. Water dilution estimates of M_{nl} were correlated ($P = 0.004$) to values calculated by difference (body mass − chemically extracted lipid mass), but the correlation was lower, and the average error higher, than for M_{nl} estimates from the gas dilution (cyclopropane) and total body electrical conductivity

(TOBEC) methods on various species (Table 4.4). The TOBEC method was more accurate than the water dilution method and approached the accuracy of the cyclopropane dilution method in estimating the M_l and M_{nl} of rabbits (Table 4.4).

Equilibration criteria

The three validations used similar but slightly different criteria for deciding when equilibrium occurred. Lesser *et al.* (1952) waited 90 to 150 minutes after injecting cyclopropane before sampling and analysing chamber gas every 30 minutes. They concluded that equilibrium had occurred when cyclopropane concentration was 'constant' for three to five consecutive samples. Similarly, Gessaman *et al.* (1998) began sampling after 150 minutes of equilibration, and continued analysing gas samples '. . . until the SD of the mean of three samples was less than 0.75%'. Henen (1991) began sampling after 3 h of equilibration and sampled every 30 min until M_l estimates, quickly calculated assuming that $W_b = 0.69$ (body mass) and $T_b = T_{chi}$, were within a 10% coefficient of variation. This last criterion utilizes the purpose of the dilution (estimating M_l) and incorporates compensation for (i) declining cyclopropane concentrations due to repeatedly withdrawing large samples for analysis, and (ii) changes in T_{ch} ($\simeq T_b$ for ectotherms) during the equilibration.

Determining the factors that limit gas equilibrations will significantly enhance our understanding and application of the gas dilution method. The minimum equilibration times for 200 to 400 g rats (1.5 to 2.5 h, Lesser *et al.*, 1952) were only about half of those for 400 to 650 g turtles (*ca.* 3 h; Henen, 1991). It is not clear what was the minimum equilibration time for pigeons (*ca.* 250 to 500 g; Gessaman *et al.*, 1998) or tortoises (*ca.* 1000 to 2400 g; Henen, 1997) although sampling in these studies began after 2.5 and 5 h, respectively. I did not attempt to determine the minimum equilibration time for the tortoises. Duplicate equilibrations at 3 to 4 d intervals produced essentially identical estimates of body composition, suggesting that the cyclopropane absorbed from the first equilibration had been eliminated from their bodies prior to the second equilibration. However, elimination rates may differ from absorption rates. From these endotherm–ectotherm contrasts, it appears that metabolic rate is not the primary determinant of equilibration time but body size may be important. Lean humans may absorb cyclopropane more slowly than more obese individuals (Lesser *et al.*, 1960) but it is unclear whether equilibration times are affected by lipid contents. Lesser *et al.* (1960) also suggested that (i) equilibration times were inversely related to equilibration system volume, (ii) gas diffusion in aqueous solutions and lipids should be rapid, and (iii) blood perfusion of tissues may affect equilibration rates. This area of research appears open.

Solubility coefficients

Using the α_l for rat fat appeared to work well for the rat, turtle and pigeon validations (Table 4.4). Furthermore, the temperature corrections for rat fat α_l (eqn. 4, Table 4.2) were applied over a broad range of T_b for turtles (23 to 31 °C; Henen, 1991) and pigeons (38.8 to 46.6 °C; Gessaman *et al.*, 1998), suggesting that the temperature corrections were at least adequate. However, similar temperature corrections are not available for α_{nl} or solubilities in water (α_w). Henen (1991) assumed that the partition coefficient of rat fat to 'fat-free lipid mass' (34.1, at 36.0 °C, Table 4.1; and eqn. 5, Table 4.2) did not vary over the range of turtle $T_{b.}$ Gessaman *et al.* (1998) reports using only one solubility coefficient for computing cyclopropane absorption in lean mass.

I used the cyclopropane solubility coefficients for water and rat fat (Table 4.1) to formulate regression equations relating solubility to temperature (Table 4.5). The equations were used to calculate the water and rat fat solubilities, and the rat fat–water partition coefficients, of cyclopropane at the temperatures used by Henen (1991) and Gessaman *et al.* (1998). For first-order approximation, assume that the water solubilities approximate those for non-lipid matter, i.e. $\alpha_w \approx \alpha_{nl}$; see Table 4.1 and Fig. 4.2. For the gas dilution calculations, Henen (1991; plus eqn. 5, Table 4.2) assumed that the lipid–non-lipid PC was constant over the range of turtle T_b (23 to 31 °C), but this is not true of the lipid–water PC (Table 4.5). The lipid–water PC for 31 and 23 °C, are 1.5 and 3.3% higher, respectively, than for 36.0 °C, the temperature for the PC assumed by Henen (1991). The lipid–water PC for 31 and 23 °C would translate to (i) α_{nl} ($\approx \alpha_w$) that are 1.5 and 3.3%, respectively, lower than that at 36.0 °C, and (ii) lipid estimates that are 0.6 and 1.3% higher (based upon the model in Table 4.3) than when assuming a constant PC, e.g. for 36.0 °C.

Gessaman *et al.* (1998) measured pigeon T_b of 38.8 to 46.6 °C and used a constant α_{nl}, the α_{nl} for 35 °C. Again, assuming that α_w gives a first-order approximation for α_{nl}, the α_{nl} for 35 °C is 9 and 29% greater than those for 38.8 and 46.6 °C, respectively (Table 4.5). This overestimation of α_{nl} would translate to 3.6 and 11.6% underestimates of lipid mass (based upon the model in Table 4.3), respectively. The PC for 38.8 and 46.6 °C are 1.4

Table 4.5. *Solubility coefficients (Bunsen, ml solute/ml solvent) and lipid–water partition coefficients for cyclopropane gas dissolved in water and rat fat (ether extract) necessary to evaluate assumptions in validations by Henen (1991) (23, 31 and 36 °C) and Gessaman et al. (1998) (35, 38.8 and 46.6 °C)*

	Solubility coefficients		
Temperature	Water	Rat fat	Partition coefficients
23	0.284	13.95	49.2
31	0.240	11.63	48.3
35	0.219	10.46	47.8
36	0.214	10.17	47.6
38.8	0.198	9.35	47.1
46.6	0.156	7.08	45.3

Notes:
Solubility and partition coefficients were calculated from regression equations derived from data in Table 4.1.[a]
[a] The regression for water solubility ($\alpha_w = 0.40741 - 0.00538 \times \text{Temp}$; $r^2 = 0.991$, $F_{1,2} = 109$, $P = 0.06$) was based upon Bunsen solubilities converted from Ostwald solubilities at 20.0, 25.0 and 37.5 °C listed in Table 4.1. The regression for rat fat solubility ($\alpha_l = 20.65 - 0.291 \times \text{Temp}$; $r^2 = 0.998$, $F_{1,2} = 542$, $P = 0.03$) used the Bunsen solubilities listed for rat fat (temperatures of 27.5, 34.0 and 37.5 °C) in Table 4.1.

and 5.3% smaller than that for 35 °C, translating roughly to 1.4 and 5.3% overestimates of α_{nl}, and 0.6 and 2.1% underestimates of lipid mass, respectively. This crude analysis helps to emphasize two points: (i) for different temperatures, it is probably more prudent to assume a constant PC than a constant α_{nl}, and (ii) it is important that α_l and α_{nl} be determined for other species, more temperatures, and non-lipid matter from different body conditions, e.g. states of dehydration, starvation or atrophy.

Solvents

The lipid extraction solvents used for rats, turtles and pigeons were ethyl ether (Lesser *et al.*, 1952), chloroform–methanol (Henen, 1991) and petroleum ether (Gessaman *et al.*, 1998), respectively. The solubility coefficients for rat fat were determined from ethyl ether extracts, so the accuracy of the turtle validation may have been partially fortuitous because a different solvent was used for extracting turtle lipids.

Soxhlet apparatuses were used for extractions in the rat and pigeon studies, while the turtle study used the Folch Method (Folch *et al.*, 1957) which does not use a Soxhlet apparatus. Ethyl ether and petroleum ether remove only non-polar lipids (Dobush *et al.*, 1985; Gessaman *et al.*, 1998), and appear to extract less non-lipid material than chloroform-methanol solvents when used in Soxhlet apparatuses (Dobush *et al.*, 1985). Unfortunately, it seems unclear how much of the non-lipid material extracted (Dobush *et al.*, 1985) was due to the solvent or due to the Soxhlet apparatus. The Folch Method was designed to extract polar and non-polar lipids for biochemical analyses. It is likely that cyclopropane is highly soluble in polar and non-polar lipids, so it would be useful to know the solubility of cyclopropane in polar lipids. This is especially true for animals that are characteristically low in body lipids, e.g. $<10\%$ of body mass (Henen, 1991), where structural or polar lipids comprise a substantial portion of the body lipids (Forbes, 1978).

If structural lipids are constant through time, relative to storage lipids, sequential lipid estimates from gas dilutions will accurately reflect changes in total lipid mass and non-polar lipid mass. Rates of lipid energy storage or use can be calculated from the changes in total lipid mass. The two-compartment model, i.e. lipid and non-lipid body mass may be inappropriate for some studies. If solubilities differ significantly between polar and non-polar lipids, the gas dilution method may help quantify a three-compartment model of the body, i.e. non-lipid, polar lipid, and non-polar lipid masses.

Suggestions for improvements and future use

The gas dilution method can accurately estimate the lipid and non-lipid masses of live animals. The complexity of the method is critical to its accuracy when applied under various body conditions, e.g. lipid contents, T_b, and hydration states. Still, the complexity of the method, plus the long equilibration periods with animals constrained inside tight chambers, may currently limit the portability of the gas dilution method and may limit its application in large numbers of animals, large animals, and animals with high metabolic rates. However, technological advances may automate most of the measurements, manipulations and calculations, making the gas dilution method simpler, more portable and quicker to implement. For instance, Sentex Systems (see Appendix) produces a portable gas chromatograph, with a computer interface that can be programmed to automatically sample and analyse chamber gas and plot the sample results. Finally, the use of other gases may improve the accuracy of the method and determine whether more detailed measurements of body composition, i.e. multicompartmental models are feasible.

REFERENCES

Behnke, A.R. (1945). The absorption and elimination of gases of the body in relation to its fat and water content. *Medicine*, **24**, 359–80.

Blumberg, A.G., LaDu, B.N., Jr., Lesser, G.T. & Steele, J.M. (1952). The determination of the solubility of cyclopropane in fats and oils with the use of the Warburg apparatus. *Journal of Pharmacology and Experimental Therapeutics*, **104**, 325–8.

Brozek, J., Grande, F., Anderson, J. T. & Keys, A. (1963). Densitometric analysis of body composition: revision of some quantitative assumptions. *Annals of the New York Academy of Science*, **110**, 113–40.

Dobush, G.R., Ankney, C.D. & Krementz, D.G. (1985). The effect of apparatus, extraction time, and solvent type on lipid extractions of snow geese. *Canadian Journal of Zoology*, **63**(8), 1917–20.

Folch, J., Lees, M. & Stanley, G.H.S. (1957). A simple method for the isolation and purification of total lipids from animal tissues. *Journal of Biological Chemistry*, **226**, 497–509.

Forbes, G.B. (1978). Body composition in adolescence. In *Human Growth*, ed. F. Falkner & J. Tanner, vol. 2, pp. 239–72. New York: Plenum.

Garn, S.M. (1963). Some pitfalls in the quantification of body composition. *Annals New York Academy of Sciences*, **110**, 171–4.

Gessaman, J.A., Nagle, R.D. & Congdon, J.D. (1998). Evaluation of the cyclopropane absorption method of measuring avian body fat. *The Auk*, **115**(1), 175–87.

Henen, B.T. (1991). Measuring the lipid content of live animals using cyclopropane gas. *American Journal of Physiology*, **261**(30), R752–9.

Henen, B.T. (1997). Seasonal and annual energy budgets of female desert tortoises (*Gopherus agassizii*). *Ecology*, **78**(1), 283–96.

Hytten, F.E., Taylor, K. & Taggart, N. (1966). Measurement of total body fat in man by absorption of ^{85}Kr. *Clinical Science*, **31**, 111–19.

Keys, A. & Brozek, J. (1953). Body fat in adult man. *Physiological Reviews*, **33**, 245–325.

Klish, W.J., Forbes, G.B., Gordon, A. & Cochran. W.J. (1984). New method for the estimation of lean body mass in infants (EMME instrument): validation in non-human models. *Journal of Pediatrics, Gastroenterology and Nutrition*, **3**, 199–204.

Lesser, G.T., Perle, W. & Steele, J.M. (1952). Measurement of total body fat in living rats by absorption of cyclopropane. *American Journal of Physiology*, **169**, 545–53.

Lesser, G.T., Perle, W. & Steele, J.M. (1960). Determination of total body fat by absorption of an inert gas: measurements and results in normal human subjects. *Journal of Clinical Investigations*, **39**, 1791–806.

Merck & Co., Inc. (1952). *The Merck Index of Chemicals and Drugs*. 6th edn. Rahway, NJ., USA: Merck and Co., Inc.

Orcutt, F.S. & Seevers, M.H. (1937). The solubility coefficients of cyclopropane for water, oils and human blood. *Journal of Pharmacology and Experimental Therapeutics*, **59**, 206 10.

Pollack, G.L. (1991). Why gases dissolve in liquids. *Science*, **251**, 1323–30.

Robbins, B.H. (1936). Studies of Cyclopropane II. Concentrations of cyclopropane required in the air and blood for anesthesia, loss of reflexes, and respiratory arrest. *Journal of Pharmacology and Experimental Therapeutics*, **58**, 251–9.

Van Slyke, D.D. (1939). Determination of solubilities of gases in liquids with use of the Van Slyke–Neill manometric apparatus for both saturation and analysis. *Journal of Biological Chemistry*, **130**, 545–54.

Appendix

Sources of materials and equipment

Many of the supplies for the gas dilution method are easily obtained from standard scientific supply companies or common chromatography supply companies. Fisher, Mallinckrodt-Baker, Sigma, Thomas and VWR (see company information below) can provide the standard chemicals needed for equilibrations, e.g. glycerol, NaOH and anhydrous $CaSO_4$ or Drierite and lipid extractions (petroleum ether, chloroform and methanol). Fisher, Thomas and VWR stock the glassware for chamber parts, e.g. desiccator lids and sleeves, glass pipettes for forming simple manometers, and three-way stopcocks (metal or Teflon).

General chromatography suppliers, Supelco or Alltech, provide the necessary supplies for gas injections and sampling, e.g. Pressure-Lok, A-2 gas-tight syringes fitted with side-port needles, and preparing standards (serum vials and Mininert valves). They also provide chamber fittings (Teflon tubing and tape; plus Teflon sheeting for the NaOH packet) and septa for the syringe adapter of the stock gas bottle. Chandler Engineering manufactures table-top chromatographs. Sentex Systems produces a portable gas chromatograph which has computer interfacing and control, allowing remote access and automated gas sampling, gas analysis and plotting of gas concentrations.

Although cyclopropane was commonly used as an anesthetic a few decades ago, its use and availability have diminished. Scott Specialty Gases supplies cyclopropane gas.

Contact information for suppliers

Alltech Associates Inc., 2051 Waukegan Road, Deerfield, IL 60015, USA (1 800 323 4321)

Chandler Engineering, PO Box 470710, Tulsa, OK 74147–0710, USA (1 918 250 7200; Fax: 1 918 459 0165)

Fisher Scientific, 585 Alpha Drive, Pittsburgh, PA 15238, USA (1 800 766 7000; 1 412 490 8300)

Mallinckrodt Baker, Inc., 222 Red School Lane, Phillipsburg, NJ 08865, USA (1 800 582 2537; 1 908 859 2151)

Scott Specialty Gases, 2600 Cajon Boulevard, San Bernardino, CA 92411, USA (1 909 887 2571)

Sentex Systems, Inc., 373 Route 46 West, Building E, Fairfield, NJ 07004, USA (1 973 439 0140)

Sigma, PO Box 14508, St. Louis, MO 63178, USA (1 800 325 3010)

Supelco, Inc., Supelco Park, Bellefonte, Pennsylvania 16823–0048, USA (1 800 247 6628, USA and Canada); 05.21.14.08 (France; Sigma-Aldrich Chimie)

Thomas Scientific, 99 High Hill Road, Swedesboro, NJ 08085-0099, USA (1 800 345 2100; 1 609 467 2000)

VWR Scientific Products, 405 Heron Drive, PO Box 626, Bridgeport, NJ 08014, USA (1 800 932 5000)

IAN SCOTT, COLIN SELMAN, P. IAN MITCHELL
AND PETER R. EVANS

5

The use of total body electrical conductivity (TOBEC) to determine body composition in vertebrates

Introduction

The use of electrical conductivity of a subject as an indirect method of assessing body composition has been made for many years, particularly in the field of biomedicine (for reviews see Van Loan, 1990; Kushner, 1992; Yanovski *et al.*, 1996). The technique employed primarily by biomedics is known as bioelectrical impedance analysis (BIA: Chapter 6), in which the electrical conductivity of the subject is measured between discrete points on the subject's surface.

BIA differs from total body electrical conductivity (TOBEC®) measurement, because, in the latter, the whole subject is placed within a chamber surrounded by a conductive coil, producing a measure of total body conductivity. The acronym TOBEC has been registered as a trademark of the main manufacturer (EM-SCAN). In theory, greater standardization can be achieved using TOBEC® than with BIA as the need to attach electrodes to the skin surface, which might cause problems in most non-human animals, is removed. The TOBEC® method of determining body composition does not cause discomfort to the subject, is relatively rapid and requires little special training to use (Fiorotto *et al.*, 1987). It has been used on a wide range of animals including fish (Bai *et al.,* 1994; Gillooly & Baylis, 1999), reptiles (Angilletta, 1999), birds, e.g. Scott *et al.* (1994, 1995) including birds' eggs (Williams *et al.*, 1997) and mammals, e.g. Gosselin & Cabanac (1996), Raffel *et al.* (1996) including humans, e.g. Presta *et. al.* (1983), Van Loan *et al.* (1987), Kretsch *et al.* (1997).

The TOBEC® technique was developed initially for the agricultural and meat processing industries, particularly for use with swine (Domermuth *et al.*, 1976) and ground meat (Harker, 1973). Subsequently, it was developed for clinical applications such as the determination of the body composition of infants (Cochran *et al.*, 1986; de Bruin *et al.*, 1995; Ryan *et al.*, 1999). Not until an evaluation of the TOBEC® method for research use with small birds and mammals was published (Walsberg, 1988), however, did interest in the technique develop within the field of ecology. Walsberg (1988) was inspired to investigate the ecological applications of the technique by the earlier work of Bracco *et al.* (1983) who utilized TOBEC® to investigate the fat content of laboratory rats.

Both Bracco *et al.* (1983) and Walsberg (1988) used the apparatus designed for the determination of the fat content of ground meat. Since their studies, there have been several changes in the design of TOBEC® instruments and in the size range available. The largest manufacturer is EM-SCAN inc. (Springfield, Illinois, USA) but other manufacturers of similar instruments exist (see below).

What is total body electrical conductivity?

Despite the differences between TOBEC® and BIA (see above), the underlying principle behind these two techniques is the same, in that they both rely on the fact that the conductivity of a living organism is determined mainly by its lean tissues. This is because these tissues contain the majority of the dissolved electrolytes. The electrical conductivity of lipids is less than 5% of that of lean tissues (Pethig, 1979). Thus a measure of an organism's conductivity can be related simply to the amount of lean mass (LM) that it contains. If the simple two-compartment model of body composition is assumed then if LM is determined, the mass of lipid can be deduced by subtracting the LM from the total body mass.

The TOBEC® technique requires that the subject is completely surrounded by a conductive coil and relies on the Harker principle (Harker, 1973). This states that, when a conductor is placed inside a solenoidal coil that is producing a time-varying electromagnetic field, energy is dissipated from the electromagnetic field and is absorbed by the conductor as a current is induced within it. The coil acts as a sensor and measures the change in energy of its electromagnetic field, which is proportional to the square of the conductive volume of the sample. Unfortunately, it is impossible to model the conductive volume accurately for an animal.

In the case of the TOBEC® technique, a coil is used that produces a 5–10 MHz electromagnetic field that varies over time, in a cylindrical space within the coil insulated from external influences. In TOBEC®

instruments, the coil surrounds a cylindrical measurement chamber into which a live animal subject is placed. The subject is a conductor and absorbs a small amount of energy (500 μW) as electrical currents are induced in it. At the frequencies at which the electromagnetic field varies in TOBEC® instruments, the majority of energy is absorbed in the subject by moderately conductive materials, e.g. hydrated lean tissues rather than highly conductive materials, e.g. metals or highly resistive materials, e.g. fur, air, fat, plastic (Anon, 1993). The measurement depends on the conductivity of intra- and extracellular fluids, which are determined by the dissolved potassium, sodium and chloride ions and by the orientation of the cell membranes, enclosing the ionic solutions, which act as dielectric barriers to current flow. Thus, while TOBEC® instruments measure the energy absorbed by the lean tissues, the values obtained are correlated with the amount of body water, as this affects the concentrations of these three ions.

To relate readings of the energy dissipated from the electromagnetic field (proportional to total body electrical conductivity) to the subject's lean mass, the instrument must be calibrated. Such calibrations are usually produced by recording TOBEC® instrument readings, often termed 'indices' for a series of subjects and then analysing the body composition of these subjects using destructive methods (for example, Walsberg, 1988; Scott *et al.*, 1991; Chapter 2). Such analyses are then used to produce regression equations that relate TOBEC® index to lean mass.

Instruments available

There are at least two manufacturers of commercially available TOBEC® instruments. These are EM-SCAN Inc. and SIGNUS® (see Appendix 1). For the analysis of body composition in wild animals, EM-SCAN retails an instrument known as the SA-3000 (Fig. 5.1), whilst SIGNUS® offers the ACAN-3 (Fig. 5.2). Both the SA-3000 and the ACAN-3 are designed to have a base detection unit, to which a range of different sizes of measurement chambers can be attached. This allows a single base unit to be used for measurements of a variety of animal species with differing size ranges. The cost of the instrument varies depending on the size of the detection chamber selected and whether or not field portability is required.

Costs for the SA-3000 (at time of press) range from approximately $8700 for the smallest sets of apparatus to $11000 for the largest. The

Fig. 5.1. The EM-SCAN, SA-3000 small animal body composition analyser, shown alongside a laptop computer.

SA-3000 is capable of measuring animals that range in size from approximately 5 g to 8000 g, although, if a user wished to work across this entire size range, the whole range of measurement chambers would need to be purchased, leading to considerably greater expense (approx. $28 000). EM-SCAN also produce laboratory-based TOBEC® instruments (e.g. HP-2) for use with humans, particularly studies investigating fat and lean mass of infants, e.g. Fiorotto *et al.* (1994), de Bruin *et al.* (1995). The manufacturers of TOBEC® instruments supply plates and holders to aid with the correct positioning and repeatability of positioning of subjects. Costs for the ACAN-3 range from approximately $4300–4500 depending on the internal diameter of the measuring chamber. The standard ACAN-3 measurement chamber is capable of holding subjects between 15–45 g, but chambers that accommodate animals of greater sizes can be made to order.

Both the SA-3000 and the ACAN-3 are portable in the field. The ACAN-3 weighs 3.6 kg and comes equipped with an internal (rechargeable battery) power supply, which allows 10 hours of continuous use. The ACAN-3 base unit has dimensions of 13×21×19.5 cm, and a standard measurement chamber of 13×13×19.5 cm. To use the SA-3000 in the field, a separate portable power pack is required. Three different-sized power packs are supplied by EM-SCAN, weighing between 3.2 and 5.9 kg. They provide between 15 and 30 hours (depending on size) of operation time. The SA-3000 base unit weighs 3.7 kg and the detection chambers range between 1.2 kg and 15.9 kg. The total weight of the apparatus required to operate the SA-3000 in the field is therefore considerably greater than that for the ACAN-3. As an alternative to a portable power pack, Scott *et al.*

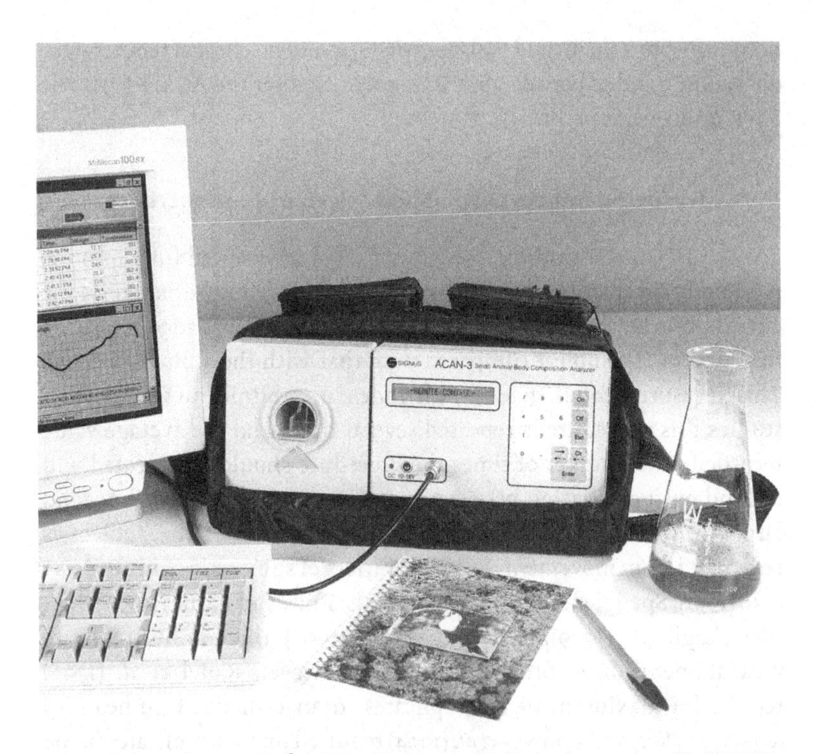

Fig. 5.2. The SIGNUS® ACAN-3 body composition analyser, a portable TOBEC instrument, shown in its carrying bag alongside a PC, through which the apparatus can be controlled and data displayed.

(1991) used an AC/DC converter to utilize a 12-volt car accumulator. It should be remembered that larger subjects require a correspondingly larger chamber, and hence instrument portability decreases, the larger the animal being worked with. When using instruments to measure TOBEC® indices in the field, it is generally the bulk of the apparatus rather than the weight that is problematic. When working with larger animals (200 g plus), it is likely that research teams will require an extra member simply to carry the apparatus.

The SA-3000 can be linked to a laptop computer to log data, but the manufacturers recommend that, in the field, data should be recorded manually from the handheld control module. The ACAN-3 comes with an internal data-logging system. TOBEC® instruments can be operated across a wide range of air temperatures. The ACAN-3 has an operating range of −5 to +40 °C, and Walsberg (1988) found no effect of temperature on TOBEC® index between 0 °C and 40 °C. Neither the ACAN-3 nor the SA-3000 are waterproof.

Measuring indices of total body electrical conductivity

To take a TOBEC® reading, the subject must be inserted fully into the measurement chamber. The instrument provides a measurement of the energy absorbed by the subject, based on the coil's impedance before the insertion of the subject compared with that with the animal inserted. This measurement is given as an index in arbitrary units. In most studies this procedure is repeated several times and the average value recorded. The number of times the procedure should be repeated will depend on the variation between repeat measures, which will depend largely on the species being studied and how well the subject is restrained from movement. In a wide variety of subjects, e.g. birds: Scott *et al.* (1991), Spengler *et al.* (1995), mammals: Buck and Barnes (1999), reptiles: Angilletta (1999), fish: Fischer *et al.* (1996), three to five replicate measurements have proved sufficient. However, Raffel *et al.* (1996) required a maximum of ten replicates in anaesthetized guinea-pigs (*Cavia porcellus*) and Conway *et al.* (1994) required up to 16 replicates in the Wood thrush (*Hylocichla mustelina*).

Positioning of the subject

Some of the older models of apparatus available to measure total body electrical conductivity, e.g. EM-SCAN SA-1 do not produce a uniform electromagnetic field. For such instruments, the most uniform section of the magnetic field occurs in the centre of the sampling area (Walsberg, 1988). It is consequently very important that subject animals are placed within the coil using a method that ensures repeatability of positioning (Stenger & Bielajew, 1995). The problem of lack of electromagnetic field uniformity may have been overcome in the ACAN-3 instrument, which the manufacturers claim produces an extremely uniform electromagnetic field (Anon, 1994). The latest apparatus developed by EM-SCAN, the SA-3000, also has a considerably more uniform electromagnetic field than their previous instruments, e.g. the SA-2. Nevertheless, neither set of apparatus has a truly uniform field. The TOBEC® reading obtained from a particular subject represents the sum of the electrical conductivity of each part of that subject's body. The contribution from any one 'slice' of body is determined by the amount of conductive material in that slice, the specific dielectric property of that conductive material, the cross-sectional area of that slice and the electromagnetic field strength at the position in the measurement chamber that the particular slice occupies (Anon, 1994). Thus the position of the whole animal in the chamber is crucial.

Repeatability with respect to the positioning of the subject, however, is also important because the interaction between the coil and the conductor (subject) is a function not only of the amount of electrolytes present but also of the cross-sectional area of the conductor (Harker, 1973). Thus, for a live subject, the reading from the apparatus will be determined by the mass of the lean tissues, and by how they are distributed throughout the coil (Anon, 1991; 1994; Fiorotto *et al.*, 1987). It should also be noted that different tissues have different abilities to conduct electricity. These tissues vary in their anatomical distribution and cross-sectional area from species to species.

Owing to the complexities of body geometry and the distribution of tissues within them, TOBEC® instruments thus give readings that are proportional to the fresh lean mass of a species of only a particular size and shape. Two subjects that are of equal body mass and total lean mass can nevertheless yield different readings if the distribution of their lean tissues is different, as can be the case with animals of different species, or individuals within a species.

Given that positioning of the subject is a vital aspect of the TOBEC® technique, care must be taken to ensure that the subject is still. In the case of birds, which have a rigid axial skeleton, this can be achieved by placing the subject into a nylon mesh stocking (used by Walsberg, 1988; Roby, 1991), restraining it with rubber bands (used by Skagen *et al.*, 1993; Conway *et al.*, 1994) or in a soft plastic cylindrical jacket with Velcro fastenings (used by Scott *et al.*, 1991; Frawley *et al.*, 1999). In the case of mammals, which tend to more flexible and more adept at escaping, the use of an anaesthetic may be required (see Walsberg, 1988; Fiorotto *et al.*, 1987), but care must be taken since TOBEC® readings from anaesthetized animals may be different from those of unanaesthetized individuals if body temperature falls (Van Loan & Mayclin, 1987; Tobin & Finegood, 1995; see below).

It should be remembered at all times by those using the TOBEC® technique that these instruments are conductivity meters and, as such, anything that can alter the ability of the subject to conduct electricity is likely to alter the TOBEC® indices that are obtained. Such factors include the hydration state of the subject, the temperature of the subject, the use of anaesthesia and the presence of metallic objects on or inside the subject – such as bands, implanted transponders or telemetry devices (Scott *et al.*, 1991).

Hydration state of the subject

Although birds and mammals tend to maintain their electrolyte concentration within relatively narrow limits (Walsberg, 1988), any factors that lead to an imbalance in electrolyte concentration will affect the TOBEC® index. Walsberg (1988) showed that re-hydration in Gambel's quail (*Callipepla gambelii*) was associated with an increase in the TOBEC® index of 15%. The measurement of conductivity indices of subjects that are likely to be dehydrated should be avoided unless the extent of dehydration can be quantified and the resulting effect on the TOBEC® index is thoroughly investigated. Animals are likely to become dehydrated either if they are ill, or have been deprived of food and water for several hours as a result of an experimental manipulation, being caught in a live trap, or being held in a holding cage following capture.

Conversely TOBEC® indices should not be measured in animals with damp or wet fur or feathers, as the external water will cause an elevation in the TOBEC® index by as much as 30%, particularly in the case of saltwater contamination (Scott *et al.*, 1991). Such subjects should be dried thoroughly before measurements are taken. If animals contaminate the measurement chamber with urine, faeces and stomach contents, these substances should be removed and the chamber cleaned before any further measurements are taken.

Temperature of the subject

The conductivity of electrolyte solutions increases with increasing temperature. Thus, the TOBEC® index is positively correlated with subject core body temperature. Scott *et al.* (1991) found the TOBEC® index of heated dead birds to rise by approximately 1.5% for each 1 °C increase in core temperature, similar to the 1.2% increase per 1 °C rise found by Walsberg (1988) in heated dead laboratory rats. The implications of these findings are that calibration graphs for TOBEC® equipment should be established for particular species in known physiological states if scatter of points around the calibration line is to be minimized. Seasonal increases in body temperature, e.g. during moult, therefore could affect the tightness of a calibration.

Castro *et al.* (1990) found that the slope of the regression line of LM against TOBEC® index differed significantly between samples of live birds and those of dead birds (heated to approximately 20 °C). The TOBEC® of a live animal is significantly different from that of a dead conspecific with the same body composition and heated to the same core temperature because surface temperature distributions are not uniform in live animals and the dielectric properties of live and dead tissues also differ. Therefore, dead animals cannot be used to calibrate TOBEC® instruments if these calibrations are subsequently to be used to measure live subjects.

However, if the aim of the calibration curve is to predict the body composition of other dead conspecifics, there appears to be no reason why a TOBEC® instrument cannot be calibrated using dead animals, as long as the temperature of each is the same. We know of one such study, where TOBEC® is being used to estimate the fresh lean mass of dead Hooded crows (*Corvus corone cornix*) (G. Malacarne, *personal communication*). These crows are being sacrificed as part of a culling programme, and the use of the TOBEC® technique obviates the need for time-consuming destructive analysis of large numbers of individuals to investigate their body composition.

Anaesthesia

Anaesthesia has been used during TOBEC® measurements in many studies (Meijer *et al.*, 1994; Schoech, 1996; Raffel *et al.*, 1996; Witter & Goldsmith, 1997; Voltura & Wunder, 1998; Buck & Barnes, 1999; Gillooly & Baylis, 1999), to ensure that all subjects are positioned uniformly in the instrument, reduce movement and possibly also reduce stress on the subjects during measurement. General anaesthesia may, however, lead to a lowering of body temperature and thus decrease the TOBEC® index for a given lean mass, as described above. Tobin and Finegood (1995) reported that the decrease in TOBEC® index recorded in laboratory rats between 4 and 44 minutes post-anaesthesia led to a consequent increase in the estimation of body fat by 63%. They suggested that laboratories using TOBEC® and anaesthesia should attempt to standardize experimental protocols to account for the effects of the anaesthesia. There are less likely to be problems when using TOBEC® to determine body composition in anaesthetised ectotherms such as fish, since the core temperature of ectotherms is more affected by external temperature than by anaesthesia (Gillooly & Baylis, 1999).

Metallic objects

The frequency of the time-varying electromagnetic field of the TOBEC® coil is set by the manufacturers to dictate that energy absorption by highly conductive materials such as metals is negligible. However, EM-SCAN advise caution when using metallic identification tags, e.g. leg bands on birds, ear tags on small mammals, fin tags on fish, since '*intermittent connections between links and tissues can stimulate a somewhat conductive structure*' (Anon, 1993). Scott *et al.* (1991) found that the presence of nickel alloy leg bands issued by the British Trust for Ornithology increased the TOBEC® index by between 13 and 45% depending on the size of the ring. In contrast, the aluminium leg bands issued by the US Fish and Wildlife Service had no significant effect on TOBEC® index (Castro *et al.*, 1990; Skagen *et al.*, 1993). It is important that workers are aware of this potential problem and determine whether or not their markers significantly affect the TOBEC® index. If TOBEC® is to be measured repeatedly in marked individuals, the mark used should either (i) not affect TOBEC® readings or (ii) be easily removed before the measurements are taken.

Osborne *et al.* (1997) reported that the presence of external radiotags on subjects did not alter TOBEC® measurements significantly. However, the wide variety of internal and external radio tags available commercially necessitates that researchers should ensure that the specific type that they are using does not affect TOBEC® readings.

Some animals may carry significant amounts of lead shot (Strandgaard, 1993; Sanderson *et al.*, 1998). This is particularly the case where the subject being studied is a quarry species. Scott *et al.* (1991) found only a small effect of lead shot on TOBEC® index but did not quantify their findings.[1] Since the use of lead is being phased out in many countries, there is a need to check the possible effects of replacements, such as steel and bismuth shot, on TOBEC® Indices.

Gut contents

It has been shown in several studies that gut contents behave in terms of energy absorption in a manner indistinguishable from other fat-free body components (Voltura & Wunder, 1998) and therefore have an effect on the TOBEC® readings obtained (Skagen *et al.*, 1993; Bachman, 1994). However, the decision whether to fast a study species, or not, prior to taking a TOBEC® measurement should not be important if experimental consistency is addressed throughout that particular study (Bachman,

1994). Waiting for complete gut voiding to take place, particularly in smaller animals, may in turn lead to problems with dehydration and its deleterious effect on TOBEC® accuracy and repeatability.

Calibration methods

Had it been possible, it would have been preferable to predict body composition from measurements of TOBEC® indices using models derived from the theory relating the conductivity of a subject to its absorption of energy when placed in the TOBEC®'s electromagnetic field. Such models should apply, in theory, equally to all species and populations under investigation. However, the conductivity, and hence absorption properties of the LM is highly dependant on the geometry of the LM and its electrically conductive compartments in relation to the direction of current flow in the coil. As stated by TOBEC® manufacturers EM-SCAN (see Anon, 1993) '*the geometry of the conductive space is difficult to model mathematically*'. To date, all predictive equations that have been used to relate TOBEC® index to LM have been empirically derived, with most using chemical analysis as a reference method. Most authors have not attempted to base their empirically derived equations on theory, but have utilized a best-fit modelling approach. However, several authors have included a parameter that relates to the volume of the conductive material, e.g. subject length or subject length cubed.

Empirically derived equations have the significant drawback that they are often group specific, that is, each calibration equation applies only to the group from which it was derived. In this context, 'group' equates to phenotypically similar individuals whose body geometry does not differ significantly from those used to provide the calibration. Individuals that progress through a series of physiological states that lead to pronounced changes in the geometry of their LM may need to be assigned to different groups at different ages and different stages of their annual life cycle.

Using TOBEC® to predict lean mass (LM)

To relate readings of total body electrical conductivity to the subject's LM, regression equations are produced for a series of subjects, incorporating their TOBEC® measurements and their LM, as determined by chemical analysis (see Walsberg, 1988; Scott *et al.*, 1991; Chapter 2). Where calibrations are performed using subjects with a wide range of total lean mass

(>100 g for the EM-SCAN SA-1 but about 30 g for the ACAN-3, the best descriptive equations produced tend to be second-order polynomial with respect to interspecific calibration, but linear with respect to intraspecific calibration curves (Walsberg, 1988; Scott *et al.*, 1991; Zuercher *et al.*, 1997). TOBEC® indices have been shown to be positively correlated with total water content (Jaramillo *et al.*, 1994) and can therefore be used to predict the water content of individuals.

Single species calibration curves have been shown to be more accurate than multi-species equations at predicting total lean mass (Scott *et al.*, 1991; Asch & Roby, 1995). Scott *et al.*, (1991) tested interspecific and species-specific calibration equations using an independent sample of five starlings (*Sturnus vulgaris*). They found that their species-specific equation for starlings, derived from birds with LMs in the range 66–85 g, could predict LM with a mean error of 0.90 g (1.2%) whilst their interspecific curve, which encompassed a LM range of 35–150 g, gave a mean error of 1.6 g (2.1%). When the same birds were tested against Walsberg's (1988) interspecific equation for birds, a curve based on a wider range of lean mass (20–170 g), a much larger mean error of 7.4 g (9.9%) was produced. We thus concur with Asch and Roby (1995) and Spengler *et al.* (1995), who suggest that a new calibration curve should be determined for each species being studied using TOBEC®.

The difference in accuracy of prediction between intra- and interspecific calibrations is predominantly due to the fact that interspecific calibration curves tend to be second-order polynomial, whereas intraspecific curves tend to be linear. This is because there is usually a larger range of LM in the interspecific calibrations and the slope of the relationship between LM and TOBEC® decreases with increasing LM over large LM ranges. Thus, the prediction of the lean mass of larger animals will be less affected by errors in TOBEC® index measurement than the LM of smaller animals when using the same interspecific calibration curve.

The second-order polynomial relationship found when the fresh lean mass range is large probably reflects the interaction between the TOBEC® instrument's coil diameter and the total cross-sectional area of the subject. Smaller animals produce less interaction with the coil and therefore produce less change in a large coil's impedance (Anon, 1994). To overcome this problem, manufacturers have produced instruments that can accept a variety of different coil/ sample chamber sizes. It is now recommended that a coil size is used such that a sample should fill at least one-half of the sample chamber that is being deployed (Anon, 1994).

Roby (1991) found no difference between the TOBEC® indices of the same subjects measured with two different sets of apparatus (EM-SCAN, SA-1). It thus seems likely that calibration curves are not specific to individual sets of apparatus, but it is highly likely that calibration curves devised for a particular type or generation of apparatus will be specific to that model.

When producing calibration curves several workers, e.g. Skagen *et al.*, (1993), Conway *et al.* (1994), Voltura & Wunder (1998), have included additional predictive biometric variables, such as culmen and tarsus length. There is a strong argument for the inclusion of such variables, given that one of the key elements in determining TOBEC® index is the total cross-sectional area of the subject, and thus any measures that explain variation in the cross-sectional area between subjects should enhance the predictive qualities of a calibration equation (Fiorotto *et al.*, 1987). The incorporation of such biometrics into equations to predict total lean mass using TOBEC® has been found to enhance their accuracy by several authors, e.g. Fiorotto *et al.* (1987), Skagen *et al.* (1993), Lyons and Haig (1995).

It is important also to note that the aspiration of those attempting to use non-invasive techniques is often to measure changes over time in body composition of the same individual. It is impossible to produce calibration curves for intra-individual changes if destructive analysis is used as the 'gold standard'. Thus, in such studies, intraspecific equations have to be used in place of intraindividual calibration. Given that geometry of the subject can impact upon its TOBEC® index, it is possible that using an intraspecific equation to follow individuals could lead to errors in predicting fresh lean mass that are difficult to quantify. Some researchers have suggested that this indicates that the TOBEC® method is not useful in following changes in body composition of an individual over time, but we think this is an unduly pessimistic view. Most studies of seasonal changes in lean tissues of birds, by destructive analysis, have indicated that these changes occur chiefly in digestive organs and other tissues inside the body cavity (Piersma *et al.*, 1993; Mitchell, 1996; Selman, 1998) and so are unlikely to affect the geometry of that individual. Any differences in the geometry of lean mass between successive measurements of the same individual are likely to be smaller than those measurements taken amongst individuals of the same species.

There has been some debate as to which variable should be used as the independent, when producing calibration curves, given that the TOBEC®

index will eventually be used as a predictor in any calibration equation. Presta *et al.* (1983), Van Loan and Mayclin (1987), Cochran *et al.* (1989) and Scott *et al.* (1991) all favoured the use of the TOBEC® index as the independent variable, whereas others, such as Walsberg (1988), Castro *et al.* (1990) and Asch and Roby (1995) prefer the use of LM as the independent variable and then convert the resultant equation algebraically to predict LM from TOBEC® index. Strictly speaking, the TOBEC® index is dependent on the lean mass of the subject and is not measured without error. Measurement of the independent variable without error is a requirement of the least-squares regression method (Sokal & Rohlf, 1995). Thus there is a good statistical reason to treat the TOBEC® index as the dependent variable. It is also the case, however, that the TOBEC® index of a subject can be recorded with high reproducibility, if care is taken with regard to positioning of the subject. It is also possible that the same TOBEC® index can be produced from two different individuals with differing fresh lean masses, if, for example, they are of slightly different body shapes or have different proportions of tissues with different dielectric properties. In addition, it is often assumed that there is little error produced when determining body composition destructively, but this is only the case when extreme care has been taken in the carcass analysis. Thus we prefer to predict LM directly from the TOBEC® index. In cases where there is the probability of both the dependent and independent variables being measured with error, it is best to deploy a type II regression model (Reduced major axis: Sokal & Rohlf, 1995). In several cases, researchers have chosen to introduce secondary variables into multiple regression-based calibration equations, in an attempt to increase the accuracy of prediction, e.g. Fiorotto *et al.* (1987), Scott *et al.* (1991), Skagen *et al.* (1993). This causes difficulties when using type II regression as computational programmes that perform multiple regression using the type II model are not widely available.

Debates concerning the appropriate regression model are, to some extent, of importance only when a functional significance is to be ascribed to the slope of the regression line. When such slopes are used only for the purposes of prediction, it is the ability of the regression equation or slope to predict accurately that determines how effective TOBEC® is as a technique.

Using TOBEC® to predict fat mass (FM)

There has been considerable debate on the indirect use of TOBEC® to estimate fat mass in birds. Early studies (Castro *et al.*, 1990; Roby, 1991; Scott *et al.*, 1991) derived fat mass (FM) by the deduction of predicted lean mass (PLM) (derived from regression models relating TOBEC® index to LM) from body mass (BM) (measured by weighing). Subsequent studies (Morton *et al.*, 1991; Skagen *et al.*, 1993) pointed out that the absolute error in PLM and predicted fat mass (PFM) must be the same but is usually a greater proportion of the FM since LM usually exceeds FM. They recommended the use of multiple regressions to estimate FM from BM and TOBEC®, since the error associated with deriving FM is independent of that associated with predicting LM.

Several studies have used multiple regression to incorporate BM, TOBEC® and various body size measurements to predict FM (Skagen *et al.*, 1993; Conway *et al.*, 1994; Meijer *et al.*, 1994; Voltura & Wunder, 1998; Frawley *et al.*, 1999). They used stepwise multiple regression to extract independent variables according to their contribution to overall variance in FM and concluded that the effectiveness of TOBEC® could be assessed on the basis of whether or not it improved significantly the multiple regression model of FM based on BM, a fat score index and various body measurements.

However, in three separate studies on three different species of shorebird, redshank (*Tringa totanus*), sanderling (*Calidris alba*) and knot (*Calidris canutus*), estimating FM by subtraction of PLM (predicted from a simple linear regression equation) from BM, yielded smaller errors than a multiple regression equation with TOBEC® index and BM as independent variables (Mitchell, 1996; Hole, 1997; Selman, 1998).

Is TOBEC® effective and useful?

Since Walsberg (1988) first suggested TOBEC® as a technique for measuring body composition in live animals, 18 published studies have employed TOBEC® to estimate body composition in laboratory animals (rats and guinea pigs), 35 in wild animals (birds, mammals, reptiles and fish) and 7 in livestock (chickens, pigs and sheep).

Table 5.1 shows that TOBEC® has been used widely to estimate body composition in laboratory animals and in clinical studies of humans. TOBEC® has proved a very reliable method for determining body composition in human infants, despite calibration curves being produced using mini-pigs. For example, de Bruin *et al.* (1998) used TOBEC® to measure the energy deposition in breast-fed and formula-fed infants and found that the current recommended guidelines for energy requirements in

Table 5.1. *Number of published studies between 1988 and 1999 (not including meeting abstracts) that have used TOBEC®*

Species studied	Number of published studies	Did not include a validation of TOBEC®	Species-specific validation of TOBEC®	[a]Positive outcome	[b]Negative outcome
Birds	**23**	**7**	[c]**16**	**9**	**7**
Wild	21	7	14		
Livestock	1		1		
Eggs	1		1		
Mammals	**30**	**14**	**16**	**10**	**6**
Laboratory	18	8	10	6	4
Livestock	6	4	2	1	1
Wild	6	2	4	3	1
Reptiles	**1**		**1**	**1**	–
Fish	**6**	**1**	**5**	**4**	**1**
Humans	**30**	**30**	–	–	–
Total	**90**	**52**	**38**	**24**	**14**

Notes:

[a] Positive outcome – TOBEC® is *more* effective than alternatives, e.g. allometric techniques, or fat scoring in predicting body composition.

[b] Negative outcome – TOBEC® is no more effective than alternatives in predicting body composition.

[c] One study carried out on birds' eggs (Williams *et al.*, 1997).

infant humans appeared to overestimate empirically derived levels. TOBEC® has also been used to construct reference centiles for standard total lean mass and fat mass in male and female infants, which have been used to identify severe nutritional and growth disorders and thus aid evaluation and efficacy of subsequent treatments (de Bruin *et al.*, 1996).

In contrast, the majority of TOBEC® studies of wild animals (70%) have concentrated on assessing the validity of the technique. However, in studies which have considered TOBEC® a valid technique for determining body composition, some hypotheses have been investigated which would prove difficult, if not impossible, to answer without the use of a non-destructive technique. In one such study, Gillooly and Baylis (1999) used TOBEC® to estimate decreases in lean mass in individual male Smallmouth bass (*Micropterus dolomieui*) during the period of parental care. They then correlated these findings with individual reproductive success. TOBEC® has also been used successfully to show that body compositional changes in individual Arctic ground squirrels (*Spermophilus parryii kennicottii*) during hibernation differed with both age and sex (Buck & Barnes, 1999).

Raffel *et al.* (1996) employed TOBEC® to compare energy allocation in reproducing and non-reproducing female guinea pigs (*Cavia porcellus*). Such chronological changes in body composition could be measured using destructive analysis of samples of animals sacrificed at successive time intervals. However, in Raffel *et al.*'s study around 100 reproducing females, 80 non-reproducing females and over 300 pups would have had to be sacrificed and destructively analysed to produce an amount of data similar to that obtained by using TOBEC® to follow changes in a much smaller number of living individuals. In this study, the initial calibration of TOBEC® required only 12 adults and 21 pups to be sacrificed. TOBEC® could also potentially be used as a labour-saving device in the study of energy allocation during avian reproduction. Williams *et al.* (1997) found that TOBEC® gave a rapid, non-destructive and accurate estimate of egg composition that could potentially lead to direct analysis of egg quality, subsequent chick growth and survival.

One instance where TOBEC® could prove advantageous over destructive analysis is in the study of body composition of rare and endangered species. In this situation it is obviously unacceptable to sacrifice large numbers of animals. Schoech (1996) examined the effect of supplemental food on body condition in the threatened Florida scrub-jay (*Aphelocoma coerulescens*), where it was impossible to sacrifice any individual birds, even to calibrate TOBEC® measurements. Thus, Schoech (1996) constructed a calibration curve using a sample of the congeneric Western scrub-jay (*Aphelocoma californica*). It could be argued that such an approach is invalid since Schoech had no way of knowing whether the geometry of the lean mass and thus energy absorption of Western scrub-jays was the same as those of Florida scrub-jays. However, his study was restricted to comparing the fat masses of conspecifics, so that relative differences in body composition could be detected despite potential errors in estimating the actual magnitude of fat mass. Furthermore, the higher fat mass detected by TOBEC® in Florida scrub-jays that received supplemental food was consistent with observations of their earlier clutch initiation and their larger clutches when compared to conspecifics that were not given food supplements.

Despite the potential advantages of using TOBEC®, its validity still remains under examination. Of the 23 validation studies carried out on wild animals, 14 have endorsed the technique as effective, while 9 have found that TOBEC® is no more effective than simpler and less expensive alternatives, e.g. allometric techniques, fat scoring. So why then has there been so much debate as to whether TOBEC® is an effective method of measuring body composition in wild animals?

The majority of studies on wild animals (Table 5.2) have been primarily interested in using TOBEC® to accurately estimate the mass of lipid carried by an individual animal, e.g. Skagen *et al.* (1993), Voltura & Wunder (1998), Frawley *et al.* (1999). However, it should be emphasized that TOBEC® does not measure lipid mass directly, but is directly proportional to the lean mass (LM). The apparent misnomer that TOBEC® instruments are 'fat meters' may have misled some researchers and raised their expectations. Most studies which did not find TOBEC® to be an effective tool to measure body composition, did find strong relationships between TOBEC® and LM but found predictions of lipid mass to be poor, e.g. Lyons & Haig (1995), Danicke *et al.* (1997). Thus it should be established from the outset what TOBEC® is being used to measure and whether or not this is appropriate to the questions being asked.

Once the objective of using TOBEC® as a tool has been determined, one then has to determine how effective it is at meeting that objective. There has been considerable variation in how researchers investigating the effectiveness of TOBEC® have interpreted the predictive power of the correlation equations they generated (see Table 5.2). Some studies have concluded TOBEC® to be effective on the basis that TOBEC® indices have been strongly correlated with the particular measurements of body condition they are being used to predict, e.g. lean mass, total body water, lean dry mass, see Roby (1991), Schoech (1996). However, regression coefficients and confidence intervals should be treated only as descriptors of the statistical relationship between the TOBEC® index and measure of body condition of those subjects used in the correlation, but not as measures of the strength of the correlational equations used to predict body condition parameters in other independent subjects.

The strength of a calibration curve of LM against TOBEC® index can be determined with confidence only by testing the ability of the equation to predict total lean mass in an independent group of subjects, i.e. a group not used in the development of the original predictive equation (for further discussion see Perdeck, 1985).

Some studies have compared the TOBEC®-derived estimate of lean mass with the actual value by linear regression, e.g. Roby (1991), Scott *et al.* (1991) and concluded that TOBEC® is 'effective' if the resultant regression

Table 5.2. *Summary of TOBEC® validation studies on wild animals, including species studied and the purpose for which TOBEC® was validated, i.e. estimating fat mass (FM), lean mass (LM), total body water (TBW)*

Study	Study species (number of individuals used)	Body mass range (grams)	[a]Mode of validation	[b]Purpose (outcome: + or −)
Mammals				
Bachman (1994)	*Spermophilus beldingi* (11)	165–245	COR MVA	LM(+) FM(−)
Koteja (1996)	*Mus musculus* (20)	14–31	CV	LM(+)FM(+) TBW(+)
	Clethrionomys glareolus (14)	16–27	CV	LM(+)FM(+) TBW(+)
Voltura & Wunder, (1998)	*Microtus ochrogaster* (26)	25–59	CV MVA	LM (+) FM (−/+)
Zuercher *et al.* (1997)	*Clethrionomys rutilus* (50)	11–40	MVA	LM (+) FM (−) TBW (+)
Walsberg (1988)	*Perognathus pennicilatus* (1)			
Species pooled	*Peromyscus maniculatus* (1)			
	Dipodomys merriami (2)			
	Mesocricetus auratus (6)			
	Neotoma lepida (3)	17.5–627	COR	LM (+)
	Dipodomys deserti (2)			
	Neotoma albigula (2)			
	Rattus norwegicus (11)			
Birds				
Asch & Roby (1995)	*Passer domesticus* (35)	23–31	COR MVA	LM (+) FM (−)
	Sturnus vulgaris (63)	63–92	COR MVA	LM (+) FM (−)
Burger (1997)	*Cardinalis cardinalis* (49)	39–56	MVA IS	LM(−) FM(−)
Castro *et al.* (1990)	*Passer domesticus* (12)			
Species pooled	*Calidris pusilla*			
	Calidris fuscicollis			
	Phalaropus tricolor (5)			
	Calidris himantopus (9)	18–90	COR	LM (+)
	Charadrius vociferus (1)			
	Gallinago gallinago (4)			
	Limnodromus scolopaceous (1)			

Table 5.2 (*cont.*)

Study	Study species (number of individuals used)	Body mass range (grams)	[a]Mode of validation	[b]Purpose (outcome: +or −)
Conway *et al.* (1994)	*Hylocichla mustelina* (21)	46–58	CV MVA	LM (+) FM (−)
Frawley *et al.*, (1999)	*Colinus virginianus* (173)	151–276	CV MVA IS	FM (+/−)
Leberg *et al.* (1996)	*Hylocichla mustelina* (24)	–	CV MVA	FM (+/−)
	Catharus ustulatus (24)		CV MVA	FM (+/−)
	Piranga rubra (24)		CV MVA	FM (+/−)
	Agelaius phoeniceus		CV MVA	FM (+/−)
Lyons and Haig (1995)	*Calidris pusilla* (20)	25.4 ± 3.3(sd)	COR CV MVA	LM (+) FM (−)
	Calidris alpina (18)	53.5 ± 5.3(sd)	COR CV MVA	LM (+) FM (−)
	Limnodromus griseus (20)	95 ± 25.5(sd)	COR CV MVA	LM (+) FM (−)
Meijer *et al.* (1994)	*Sturnus vulgaris* (18)	2–104	COR MVA	LM (+)FM (−)
Morton *et al.* (1991)	*Philohela minor* (14)	106–215	COR MVA	LM (+) FM (−)
Roby (1991)	*Colinus virginianus* (62)	160–256	COR	LM (+) FM (+)
Schoech (1996)	**Aphelocoma californica* (17)	–	COR	LM (+)
Scott *et al.* (1991)	*Calidris alpina* (11)	40–60	COR	LM (+)
	Tringus totanus (6)	146–181	COR	LM (+)
	Charadrius hiaticula (6)	59–82	COR	LM (+)
	Sturnus vulgaris (10)	62–93	COR IS	LM (+)
Skagen *et al.* (1993)	*Calidris pusilla* (24)	19–24	COR CV MVA	LM (+) FM (+)
	Calidris. fuscicollis (22)	33–41	COR CV MVA	LM (+) FM (+)
Spengler *et al.* (1995)	*Hylocichla mustelina* (24)	(11–15)	MVA CV	FM (+)
	Catharus ustulatus (24)	(6.7–8.8)	MVA CV	FM (+)
	Piranga rubra (24)	(7.2–9.0)	MVA CV	FM (−)
	Agelaius phoeniceus (24)	(15–18)	MVA CV	FM (−)
Walsberg (1988)	*Carpodacus mexicanus* (3)			
Species pooled	*Zonotrichia leucophrys* (2)			
	Passer domesticus (1)			
	Cardinalis cardinalis (1)			

	Columbina inca (3)	15–170	COR	LM (+)
	Mimus polyglottus (1)			
	Melanerpes uropygialis (2)			
	Quiscalus mexicanus (4)			
	Zenaida macroura (2)			
	Zenaida asiatica (1)			
	Callipepla gambelii (7)			
Williams *et al.* (1997)	Bird egg composition			
Species pooled	Domestic chicken (25)	37.5–76.3		
	Domestic duck (15)	61.6–85.8		
	Domestic Guinea fowl (14)	40.0–51.4	COR MVA	LM (+) FM (+)
	Domestic quail (7)	10.2–13.6		
Reptiles				
Angilletta (1999)	*Sceloporus undulatus*(37)	5–15	COR MVA	LM (+) FM (+)
Fish				
Bai *et al.* (1994)	*Sciaenops ocellatus* (40)	10–138	COR MVA	LM (+) FM (+)
Fischer *et al.* (1996)	*Lepomis macrochitus* (36)	25–106	COR	LM (+) FM (+)
Gillooly & Baylis (1999)	*Micropterus dolomieui* (21)	550–1295	MVA	LM (+)
Jaramillo *et al* (1994)	*Ictalurus punctatus* (74)	44–175	COR MVA IS	LM (+) FM (+)
Lantry *et al.* (1999)	*Perca flavescens* (43)	18–434	COR MVA	TBW (−)
	Alosa pseudoharengus (47)	21–48	COR MVA	TBW (−)

Notes:

The range of body mass of those animals used to calibrate the TOBEC® with destructive analysis are shown. For those studies which did not supply body mass range information, this column is either left blank or shows mean body mass or ranges of LM in parentheses.

* Study species Florida scrub jay *Aphelocoma coerulescens* is a threatened species , therefore congeneric Western scrub-jay *A. californica* used to validate TOBEC®.

[a] CV = cross-validation; IS = independent test sample; COR = significant correlation between body composition parameter and TOBEC®; MVA = multivariate analysis including the contribution of TOBEC® to explaining variation in the body composition parameter.

[b] FM(−) = authors concluded TOBEC® to be ineffective at predicting mass of fat; LM(+) = authors concluded TOBEC® to be effective at predicting lean mass, etc. (+/−) = authors do not comment on effectiveness of TOBEC®.

line has a slope equal to one and an intercept term equal to zero. Such an approach determines if the estimate is in broad agreement with the actual value, but only by calculating the difference between the estimate and the actual value can the exact error of estimation be determined.

If, for some reason, the number of subjects available for use in calibration is in limited supply, such that it is not possible or it is considered unethical to take a second independent sample for the purposes of testing the calibration equation, a re-sampling or jack-knife procedure could be utilized (Sokal & Rohlf, 1995). Such a procedure is often referred to as 'cross-validation', e.g. Skagen *et al.* (1993), Conway *et al.* (1994). Cross-validation involves sequentially removing a single subject (datum point) from the calibration sample data set and estimating that subject's body composition using a predictive equation derived from the remaining data points. The single subject point is then replaced in the set and a different data point then removed. This is repeated for each subject in the calibration sample. Each predicted value of body composition is thereby effectively calculated independently. The differences between the predicted and actual values are then calculated to give a set of independently determined errors. However, a set of truly independent individual points to test the predictive equations is to be preferred to the procedure of cross-validation used by Skagen *et al.* (1993), although a second sample of individuals needs to be sacrificed.

Independent testing and cross-validation both yield the magnitude of the error expected from the calibration curve, i.e. the difference between the actual value and the TOBEC® estimate. Expressing errors as percentages of the quantity being estimated is often used to quantify the effectiveness of TOBEC® in the context in which it is being used (Morton *et al.*, 1991; Scott *et al.*, 1991; Skagen *et al.*, 1993; Conway *et al.*, 1994; Zuercher *et al.*, 1997). For example, Skagen *et al.* (1993) used cross-validation to calculate the errors of estimation of lean mass (LM) from intraspecific relationships between LM and TOBEC® (dependent variable) of semipalmated sandpipers (*Calidris pusilla*) (LM of 20–24 g) and white-rumped sandpipers (*Calidris fuscicollis*) (LM of 31–40 g) and found mean errors to be 3.3% ± 0.42% (s.e.) and 3.5% ± 0.52% (s.e.), respectively, of the LM being estimated. However, the usefulness of expressing errors as percentages is questionable and potentially misleading. For instance, a mean absolute error of 5 g in estimating LM would equate to a 10% error in a bird of 50 g, but only a 5% error in a 100 g bird. This would suggest that, in larger birds, the error of prediction is smaller, which is misleading since the absolute error was the same. Errors are more indicative of the accuracy of the application of a predictive model if they are expressed in absolute terms and in the context of the range over which parameters such as LM are being estimated (see examples below). If the error exceeds the range over which LM varies either between individuals or within individuals (depending on the comparisons being made), the resolution of the predictive equations and the TOBEC® technique obviously is not fine enough. If we return to the example of Skagen *et al.* (1993), the mean absolute errors in estimating LM were, in fact, 0.71 g $\pm$ 0.826 g (95% CI) in semipalmated sandpipers (LM of 20–24 g) and 1.32 g $\pm$ 1.939 g in white-rumped sandpipers (LM of 31–40 g). By expressing these errors in absolute terms, we can see that, if TOBEC® is to be used to detect (with 95% confidence), variation in LM either within individuals or between individuals, variation in LM would need to be in excess of 1.53 g in semipalmated sandpipers and 3.26 g in white-rumped sandpipers. Given that the variation in LM between individual sandpipers was only 4 g in semipalmated and 9 g in white-rumped, the error of the prediction could account for 38% and 36%, respectively, of the variation in LM in the two species. Thus the effectiveness of Skagen *et al.'s* (1993) TOBEC® equations for estimating variation in LM in the two species of sandpiper is much less than the percentage errors of 3.3% and 3.5% they originally suggested. This example illustrates how the effectiveness of TOBEC® (in predicting body composition) can be ascertained fairly only by considering the magnitude of the absolute errors of the calibration in the context of the variation in the body composition parameter being predicted.

TOBEC® and any other technique used to estimate body composition is useful only if the prediction errors associated with estimates of body composition are smaller than the variation exhibited by the parameter that is being estimated. Thus, even apparently large errors may not render the TOBEC® technique ineffective, if the variation in body composition is greater than those errors. Some studies have reported that the inclusion of the TOBEC® index does not improve the ability of equations based on body mass and other biometrics to predict fat mass (FM), e.g. Conway *et al.* (1994), Lantry *et al.* (1999). Given the importance of quoting absolute errors and of setting them in the context of the degree of variation in the parameter being estimated, it is perhaps surprising that so few studies actually quote these values. Thus it is very difficult to place quantitative bounds on the effectiveness of the TOBEC® technique in many published studies.

Despite being based on widely varying criteria, the conclusions of the validation studies listed in Table 5.2 illustrate that TOBEC® is, on the whole, more effective when used to predict body composition of larger animals. This is not surprising, given that most studies have used the earlier EM-SCAN equipment, i.e. models SA-1 or SA-2, both of which have a fixed chamber diameter of *c.* 100 mm. As suggested by Asch and Roby (1995), this chamber size may be larger than the optimum for many animals smaller than 100 g, since a subject's cross-sectional area should ideally be at least half that of the measuring chamber. The introduction of different-sized chambers will hopefully improve the effectiveness of TOBEC® techniques when applied to small animals.

It is important that future validation studies of TOBEC® adopt a consistent approach to calibration, in terms of LM error calculation and the assessment of effectiveness. Otherwise the debate on the value of TOBEC® in measuring the body composition of animals will continue unresolved. In the following section, we use an example to suggest the way in which future studies should approach the validation of TOBEC®.

The example uses totally independent data points to test the accuracy of a predictive equation. Data are presented from Selman (1998), who validated the use of TOBEC® to estimate the total lean mass and hence the fat mass of captive knot *Calidris canutus*, a migratory shorebird.

Context of study: What is the reason for using TOBEC®

It was intended to use TOBEC® to estimate repeatedly the LM and FM of individual knot held in captivity throughout a whole year. The measurements of body composition were taken in conjunction with measurements of metabolic rate, so that any seasonal variation in metabolism could be related to any parallel changes in body composition.

Calibration

TOBEC® was measured in 11 captive knot (Sample A), which were then humanely killed immediately afterwards. The actual lean masses (LM) of the carcasses were determined by chemical analysis and regressed against the corresponding TOBEC® indices. If biometrics are to be included in the calibration procedure, it is important that they are measured prior to the death of the calibration subjects. The resulting predictive linear regression equation is given below (eqn. 1, Fig. 5.3). Other curve types, e.g. second-order polynomial were also fitted to the calibration data, but proved less accurate than the linear equation and so are not given here.

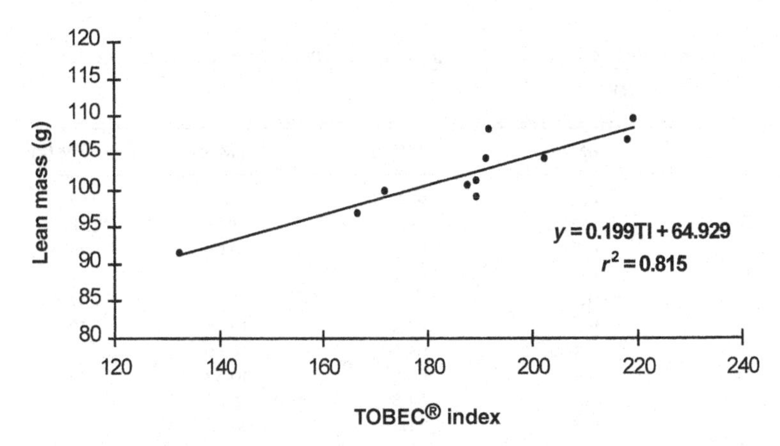

Fig. 5.3. The relationship between lean mass and TOBEC® index (TI) and LM for a group of 11 captive knot.

$$LM = (0.199^{*}TI) + 64.929 \quad (1)$$

where TI = the TOBEC® index

In addition, a multiple regression equation (eqn. 2) was calculated incorporating both body mass (BM) and TOBEC® index as independent variables to predict fat mass (PFM_m) (dependent variable) (Morton *et al.*, 1991; Skagen *et al.*, 1993). Biometrics, if recorded, could also be included in the calibration equations.

$$PFM_m = (0.982 \times BM) - (0.182 \times TI) - 65.5 \quad (2)$$

The BM, LM and FM of each individual in Sample A are given in Table 5.3. Note that the percentage water content (mass of water as a percentage of LM) of each individual is also given. It is important to measure percentage water to check that all of the individuals used to calibrate TOBEC® were of normal hydration (between 60–70% of LM) when TOBEC® was measured.

Validation

The predictive power of the TOBEC® calibration equation (eqn. 1) was quantified using a second independent sample of 9 captive knot (Sample B). TOBEC® was measured in each individual in sample B and eqns. 1 and 2 were used to calculate predicted Lean Mass PLM and PFM_m, respectively. A second estimate of fat mass PFM_s was calculated by subtracting PLM from BM (eqn. 3).

$$PFM_s = BM - PLM \quad (3)$$

Table 5.3. *Body composition data of 11 captive knot used to construct TOBEC® calibration curve*

Bird	BM (g)	%Water	LM (g)	FM (g)
A	131	67.1	108.3	22.4
B	120	65.6	91.5	28.9
C	126	66.5	104.9	20.9
D	114	66.1	103.3	12.3
E	110	64.5	96.9	12.8
F	163	63.3	106.8	56.6
G	153	65.1	100.0	52.9
H	128	63.8	100.7	26.9
I	122	64.8	99.2	23.2
J	148	64.8	109.8	38.0
K	155	64.1	104.4	50.6
Mean	134	65.1	102.2	31.4
s.e.	5.1	0.3	1.5	4.5

Errors of estimation were calculated as the differences between LM and PLM, FM and PFM_m, FM and PFM_s, for each individual in sample B (Table 5.4). Note that the absolute errors for PLM and PFM_s are necessarily identical. When calculating mean errors, the sign of each error was ignored and the modulus taken. A similar procedure should be conducted for all calibration equations produced, e.g. those including biometrics.

The mean absolute error in calculating PLM was 1.4 ± 0.7 g (95% CI). This means that the TOBEC® can detect (with 95% confidence) interindividual differences or intraindividual changes in LM of >2.1 g. The range of LM in the 9 birds in sample B was 13.4 g (91.6–104.8 g) and within individual knot, LM was found to vary by up to 12 g. Therefore, the resolution for detecting interindividual differences and intraindividual changes in LM is 15.6% and 17.5%, respectively.

The mean absolute error in estimating PFM_s was the same as for PLM and was lower than the mean error in calculating PFM_m using multiple regression (eqn. 2), which was 2.2 ± 1.4 g (95% CI). Therefore, in captive knot, fat mass is best predicted by subtracting PLM from BM. The range of FM in the 9 knot in Sample B was 45.2 g (7.2 – 52.3 g) and intraindividual changes within a season were as high as 100 g. Thus the resolution of TOBEC® for detecting interindividual differences and intraindividual changes in FM is 4.9% and 2.2%, respectively. A similar assessment

Table 5.4. *Independent comparison of the errors between the actual values of total lean mass and total fat mass (LM, FM) and the predicted values (PLM, PFM) obtained using predictive models derived from total body electrical conductivity TOBEC®*

Bird	BM(g)	Water %	LM (g)	LM-PLM (g)	FM (g)	FM-PFM_s (g)	FM-PFM_m (g)
1	109	63.0	97.6	+1.2	11.3	−1.2	−4.5
2	112	63.9	92.3	+0.3	20.0	−0.3	−3.0
3	112	63.0	98.1	−2.8	13.7	+2.8	−0.1
4	118	64.5	94.3	−1.8	23.8	+1.8	−0.9
5	112	61.9	104.8	+3.1	7.2	−3.1	−5.7
6	120	66.5	96.1	−0.5	24.2	+0.5	−1.2
7	112	64.2	91.9	+1.1	20.5	−1.1	−2.3
8	121	65.2	93.9	+2.0	26.6	−2.0	−1.7
9	144	65.5	91.6	+0.2	52.3	−0.2	−0.6
Mean	118	64.2	95.6	1.4*	22.2	1.4*	2.2*
s.e.	3.4	0.5	1.3	0.3	4.1	0.3	0.6
95% CI	±7.8	±1.2	±3.0	±0.7	±9.5	±0.7	±1.4

Notes:
PLM Derived from linear regression (eqn. 1)
PFM_s = BM − PLM
PFM_m derived from multiple regression (eqn. 2)
% water = (total water content/LM) × 100
* Mean calculated ignoring sign (+ or −) of each error.

procedure should be conducted for all calibration equations produced and the optimum chosen.

The use of total body electrical conductivity in longitudinal studies

As mentioned previously, one of the great advantages of non-destructive techniques to estimate body composition is that the same individual can be measured more than once and thus changes in body composition can be followed over time. To date, however, the majority of calibration studies have focused on groups of supposedly similar individuals. This has sometimes led to the spurious assumption that such calibrations automatically apply to other groups or individuals, at all points in their adult life and annual cycle. Such an assumption would apply only if, throughout an individual's life, the distribution and make-up of its dielectrically active tissues remained constant. Several organisms are known to show changes in the concentration of their electrolytes during their annual cycles. For example, during pregnancy, the hydration state of many animals is known to change (see Fergusson & Bradshaw, 1991; Bolton *et al.*, 1996). Many birds show changes in both body water content and body temperature during moult (see Newton, 1969). For such species it is unlikely that a single calibration curve would be sufficient to provide accurate estimates of body composition throughout the entire annual cycle, unless the calibration was based on animals in all physiological states. Such calibrations would include considerable 'noise' around the basic allometric relationship and hence, provide rather wide confidence limits for predictive purposes. Bell *et al.* (1994) assessed the ability of the TOBEC® technique to assess experimentally induced changes in the body composition of rats under a range of different conditions. They concluded that TOBEC® was inaccurate across experimental conditions, and within a single experimental condition during the course of an experiment. Unusually they did not find a direct link between TOBEC® index and LM. In contrast to Bell *et al.*'s study, Danicke *et al.* (1997) followed the growth of male broiler chickens from hatching until adulthood under a range of different dietary conditions. Despite these conditions resulting in marked differences in body weights and body chemical compositions, Danicke *et al.* found similar TOBEC® responses for a given LM. Similarly, Cochran *et al.* (1989) found that predictions based on the LM of piglets, whose extracellular fluid (ECF) had been increased by up to 34%, were accurate even though the predictions were based on calibration equations developed from the same piglets prior to their ECF volume being increased. In another study, Cunningham *et al.* (1986) found no apparent effect of the administration of a diuretic or the induction of malnutrition in Sprague–Dawley rats on the relationship between LM and TOBEC® Index. Thus, whilst there is a strong theoretical reason to suspect that change in hydration status of individuals should alter the TOBEC® indices recorded for a given LM, the empirical evidence remains equivocal.

The scope for longitudinal studies of individual animals (especially birds) in the wild, rather than in captivity, is more limited, as it may be impossible to recapture such marked animals at predetermined times of year. Two studies indicate that wild birds carry greater LMs than captive birds of the same skeletal size, and have TOBEC® indices that lie above the range used in establishing the calibration curves for captives. For knot

(Selman, 1998) and dunlin *Calidris alpina* (Al-Mansour, 2000), the calibration lines of the LMs against TOBEC® indices for wild birds do not coincide with the extrapolations from captive birds. The most plausible explanation for this is that the relative anatomical distribution of the conductive tissues within an animal alters after a period of captivity through differential atrophy of lean tissues according to the levels of disuse and in response to changes in diet. Whatever the reason, it is highly desirable to establish separate calibration curves for wild and captive individuals of a study species.

Given the studies that have been described above, when TOBEC® is used in longitudinal studies, researchers should ensure that they have calibrations appropriate to the physiological status that their subjects will encounter.

Conclusions

TOBEC® is a readily accessible technique to measure body composition that is relatively cheap and easy to deploy in field situations. Most of the studies that have attempted to validate the TOBEC® technique for use in wild vertebrates have found a strong relationship between the TOBEC® index and LM. Far fewer studies have found that the technique is capable of predicting FM (by deduction), to a satisfactory level. In most cases, this is because the absolute error in the prediction of FM is a high proportion of the overall variation of FM, within individuals or between experimental groups. Some studies have suggested that this problem can be partially overcome by using TOBEC® to follow sample groups rather than individuals, e.g. Danicke *et al.* (1997), or as a qualitative rather than a quantitative instrument. In species where absolute variation in fat mass is low, it seems unlikely that the TOBEC® technique will be satisfactory. Researchers should expect errors in the range of 2–5% of LM where appropriate calibration procedures have been followed.

Whether or not the errors obtained when using TOBEC® satisfy the required demands of a particular study depends on the specific requirements of that investigation. Thus the effectiveness of the TOBEC® technique can be judged only in the context of the particular study for which its use is intended. Care must be taken to ensure that calibration curves that are in use are appropriate for the physiological state of the species under investigation.

Appendix

Useful addresses

Manufacturers

EM-SCAN

713 West Prospect Avenue, Springfield, IL 62704-5026, USA

SIGNUS®

Signus F.P.H.U. Bronowicka 42 lok-408, PL-30-091 Krakow, Poland

REFERENCES

Al-Mansour, M.I. (2000). Aspects of the life-cycle energetics of two subspecies of Dunlin *Calidris alpina*. Unpublished PhD thesis, University of Durham, UK.

Angilletta, M.J., Jr. (1999). Estimating body composition of lizards from total body electrical conductivity and total body water. *Copeia*, **3**, 587–95.

Anonymous (1991). *EM-SCAN/TOBEC® Model SA-3000 Small Animal Composition Analysis System. A Precise Analytical Instrument*. Springfield, IL: EM-SCAN inc.

Anonymous (1993). *EM-SCAN/ TOBEC® Small Animal Body Composition Analyzer Technical Brief.* Springfield, IL: EM-SCAN inc.

Anonymous (1994). *ACAN-2 Small Animal Body Composition Analyzer User's Manual.* Kraków-Poland: JAGMAR.

Asch, A. & Roby, D. (1995). Some factors affecting precision of the total body electrical conductivity technique for measuring body composition in live birds. *Wilson Bulletin*, **107**(2), 306–16.

Bachman, G.C. (1994). Food restriction effects on the body-composition of free-living ground-squirrels, *Spermophilus beldingi*. *Physiological Zoology*, **67**, 756–70.

Bai, S.C., Nematipour,G.R., Perera, R.P., Jaramillo, F., Murphy, B.R., & Gatlin, D.M. (1994).Total-body electrical-conductivity for non-destructive measurement of body-composition of red drum. *Progressive Fish-Culturist*, **56** (4), 232–6.

Bell, R.C., Lanou, A.J., Frongillo,E.A., Levitsky, D.A. & Campbell, T.C. (1994). Accuracy and reliability of total-body electrical-conductivity (TOBEC) for determining body-composition of rats in experimental studies. *Physiology and Behavior*, **56**(4), 767–73.

Bolton, L.M., Thomas,T.H. & Dunlop, W. (1996). Erythrocyte ion and water balance and membrane potential in the puerperium of normal pregnancy. *British Journal of Obstetrics and Gynaecology*, **103**, 547–51.

Bracco, E.F., Yang, M.U., Segal, K.R., Hashim, S.A. & van Itallie, T.B. (1983). A new method for estimation of body composition in the live rat. *Proceedings of the Society for Experimental Biology and Medicine*, **174**, 143–6.

Buck, C.L. & Barnes, B.M. (1999). Annual cycle of body composition and hibernation in free-living arctic ground squirrels. *Journal of Mammalogy*, **80** (2), 430–42.

Burger, M.F. (1997). Estimating lipid and lean masses in a wintering passerine: an evaluation of TOBEC. *The Auk*, **114**, 762–9.

Castro, G., Wunder, B.A. & Knopf, F.L. (1990). Total body electrical conductivity (TOBEC) to estimate total body fat of free-living birds. *The Condor*, **92**, 496–9.

Cochran, W.J., Klish, W.J., Wong, W.W. & Klein, P.D. (1986). Total body electrical conductivity used to determine body composition in infants. *Paediatric Research*, **20**, 561–4.

Cochran, W.J., Fiorotto M.L., Sheng, H.P. & Klish, W.J. (1989). Reliability of fat-free mass estimates derived from total-body electrical conductivity measurements as influenced by changes in extracellular fluid volume. *American Journal of Clinical Nutrition*, **49**, 29–32.

Conway, C.J., Eddleman, W.R. & Simpson, K.L. (1994). Evaluation of lipid indices of the wood thrush. *The Condor*, **96**, 783–90.

Cunningham, J.J., Molnar, J.A., Meara, P.A. & Bode, H.H. (1986). In vivo total body electrical conductivity following perturbations of body fluid compartments in rats. *Metabolism*, **35**, 572–5.

Danicke, S., Halle, I. & Jeroch, H. (1997). Evaluation of the non- invasive TOBEC (total body electrical conductivity) procedure for prediction of chemical components of male broilers with special consideration of dietary protein level. *Archives of Animal Nutrition*, **50**(2), 137–53

de Bruin, N.C., van Velthoven, K.A.M., Stijen, T., Juttmann, R.E., Degenhart, H.J. &Visser, H.K.A. (1995). Body fat and fat-free mass in infants: new and classic anthropometric indexes and prediction equations compared with total-body electrical conductivity. *American Journal of Clinical Nutrition*, **61**, 1195–205.

de Bruin, N.C., van Velthoven, K.A.M., de Ridder, M. *et al.* (1996). Standards for total body fat and fat-free mass in infants. *Archives of Disease in Childhood*, **74**, 386–99.

de Bruin, N.C., Degenhart, H.J., Gal, S., Westerterp, K.R., Stijen, T. & Visser, H.K.A. (1998). Energy utilization and growth in breast-fed and formula-fed infants measured prospectively during the first year of life. *American Journal of Clinical Nutrition*, **67**, 885–96.

Domermuth, W., Veum, T.L., Alexander, M.A., Hendrick, H.B., Clark, J. & Eklund, D. (1976). Prediction of lean body composition of live market weight swine by indirect methods. *Journal of Animal Science*, **43**, 966–76.

Fergusson, B. & Bradshaw, S. D. (1991). Plasma arginine–vasotocin, progesterone, and luteal development during pregnancy in the viviparous lizard *Tiliqua rugosa*. *General Comparative Endocrinology*, **82**, 140–51.

Fiorotto, M.L., Cochran, W.J., Funk, C. R., Hwai-Ping, S. & Klish, W.J. (1987). Total body electrical conductivity measurements: effects of body composition and geometry. *American Journal of Physiology*, **252**, R794–800.

Fiorotto, M.L., de Bruin, N.C., Brans, Y.W., Degenhart, H.J. & Visser, H.K.A. (1994). Total body electrical conductivity measurements: an evaluation of current instrumentation for infants. *Pediatric Research*, **37**, 94–100.

Fiorotto, M.L., de Bruin, N.C., Brans, Y.W., Degenhart, H.J & Visser, H.K.A. (1995). Total body electrical conductivity measurements: an evaluation of current instrumentation for infants. *Pediatric Research*, **37**, 94–100.

Fischer, R. U., Congdon, J.D. & Brock, M. (1996). Total body electrical conductivity (TOBEC): a tool to estimate lean mass and non-polar lipids of an aquatic organism. *Copeia*, **2**, 459–62.

Frawley, B.J., Osborne, D.A., Weeks, H.P., Burger, L.W. & Dailey, T.V. (1999). Use of total body electrical conductivity to predict northern bobwhite lipid mass. *Journal of Wildlife Management*, **63**(2), 695–704.

Gillooly, J.F. & Baylis, J.R. (1999). Reproductive success and the energetic cost of parental care in male smallmouth bass. *Journal of Fish Biology*, **54**, 573–84.

Gosselin, C. & Cabanac, M. (1996). Ever higher – constant rise of body-weight set-point in growing Zucker rats. *Physiology and Behavior*, **60**, (3), 817–21.

Harker, W.H. (1973). Methods and apparatus for measuring fat content in animal tissue either in vivo or in slaughtered and prepared form. US Patent 3735 247.

Hole, D.G. (1997). Ecophysiological preparation for migration in Sanderling *Calidris alba*. Unpublished MSc thesis, University of Durham, UK.

Jaramillo, F., Bai, S.C., Murphy, B.R. & Gatlin, D.M. (1994). Application of electrical-conductivity for non-destructive measurement of channel catfish, *Ictalurus-punctatus*, body-composition. *Aquatic Living Resources*, **7**(2), 87–91.

Koteja, P. (1996). The usefulness of a new TOBEC instrument (ACAN) for investigating body composition in small mammals. *Acta Theriologica*, **41**, 107–12.

Kretsch, M.J., Green, M.W., Fong, A.K.H., Elliman, N.A. & Johnson, H.L. (1997). Cognitive effects of a long-term weight reducing diet. *International Journal of Obesity*, **21**, 14–21.

Kushner, R.F. (1992). Bioelectrical impedance analysis: a review of principles and applications. *Journal of the American College of Nutrition*, **11**, 199–209.

Lantry, B.F., Stewart, D.J., Rand, P.S. & Mills, E.L. (1999). Evaluation of total body electrical conductivity to estimate whole-body water content of yellow perch, *Perca flavescens* and alewife, *Alosa pseudoharengus*. *Fishery Bulletin*, **97**(1), 71–9.

Leberg, P.L., Spengler, T.J. & Barrow, W.C. (1996). Lipid and water depletion in migrating passerines following passage over the Gulf of Mexico. *Oecologia*, **106**(1), 1–7.

Lyons, J.E. & Haig, S.M. (1995). Estimation of lean and lipid mass in shorebirds using total body electrical conductivity. *The Auk*, **112**(3), 590–602.

Meijer, T., Mohring, F.J. & Trillmich, F. (1994). Annual and daily variation in body mass and fat of Starlings *Sturnus vulgaris*. *Journal of Avian Biology*, **25**, 98–104.

Mitchell, P.I. (1996). Energetics and nutrition of British and Icelandic Redshank (*Tringa totanus*) during the non- breeding season. Unpublished PhD Thesis, University of Durham

Morton, J.M., Kirkpatrick, R.L. & Smith, E.P. (1991). Comments of estimating total body lipids from measures of lean mass. *The Condor*, **93**, 463–5.

Newton, I. (1969). Winter fattening in the bullfinch. *Physiological Zoology*, **42**, 96–107.

Osborne, D.A., Frawley, B.J. & Weeks Jr., H.P. (1997). Effects of radio tags on captive northern bobwhite (*Colinus virginianus*) body composition and survival. *American Midland Naturalist*, **137**, 213–24.

Perdeck, A.C. (1985). Methods of predicting fat reserves in the Coot. *Ardea*, **73**, 139–46.

Pethig , R. (1979). *Dielectric and Electrical Properties of Biological Materials*. Chichester: John Wiley.

Piersma, T., Koolhass, A. & Dekinga, A. (1993). Interactions between stomach structure and diet choice in shorebirds. *Auk*, **110**(3), 552–64.

Presta, E., Wang, J., Harrison, G.G., Bjorntop, P., Harker, W.H. & Van Itallie, T.B. (1983). Measurement of total body electrical conductivity: a new method for estimation of body composition. *American Journal of Clinical Nutrition*, **37**, 735–9.

Raffel, M., Trillmich, F. & Honer, A. (1996). Energy allocation in reproducing and non-reproducing guinea-pig (*Cavia porcellus*) females and young under *ad-libitum* conditions. *Journal of Zoology*, **239**, 437–52.

Roby, D.D. (1991). A Comparison of two non-invasive techniques to measure total body lipid in live birds. *The Auk*, **108**, 509–18.

Ryan, A.S., Montalto, M.B., Grohwargo, S. *et al.* (1999). Effects of DHA-containing formula on growth of preterm infants to 59 weeks postmenstrual age. *American Journal of Human Biology*, **11**, 457–67.

Sanderson, G.C., Anderson, W.L., Foley, G.L. *et al.* (1998). Effects of lead, iron, and bismuth alloy shot embedded in the breast muscles of game-farm mallards. *Journal of Wildlife Diseases*, **34**(4), 688–97.

Schoech, S.J. (1996). The effect of supplemental food on body condition and the timing of reproduction in a cooperative breeder, the florida scrub-jay. *The Condor*, **98**, 234–44.

Scott, I., Grant, M. & Evans, P.R. (1991). Estimation of fat-free mass of live birds: use of total body electrical conductivity (TOBEC) measurements in studies of single species in the field. *Functional Ecology*, **5**, 314–20.

Scott, I., Mitchell, P.I. & Evans, P.R. (1995). The reliability of fat scores as predictors of the mass of fat carried by individual birds. *Ardea*, **83**, 359–63.

Scott, I., Mitchell, P.I. & Evans, P.R. (1996). How does variation in body composition affect the basal metabolic rate of birds? *Functional Ecology*, **10**, 307–13.

Selman, C. (1998). The causes of individual and seasonal variation in the metabolic rate of Knot *Calidris canutus*. Unpublished PhD Thesis, University of Durham.

Skagen, S.K., Knopf, F.L. & Cade, B.S. (1993). Estimation of lipids and lean mass of migrating sandpipers. *The Condor*, **95**, 944–56.

Sokal, R.R. & Rohlf, F.J. (1995). *Biometry*. 3rd. edn, New York: W.H. Freeman.

Spengler, T.J., Leberg, P.L. & Barrow, W.C. (1995). Comparison of condition indexes in migratory passerines at a stopover site in coastal Louisiana. *Condor*, **97**(2), 438–44.

Stenger, J. & Bielajew, C. (1995). Comparison of TOBEC-derived total-body fat with fat pad weights. *Physiology and Behaviour*, **57**(2), 319–23.

Strandgaard, H. (1993). Investigation on the lethal effects of iron versus lead shot. *Zeitschrift für Jagdwissenschaften*, **39**(1), 34–45.

Tobin, B.W. & Finegood, D.T. (1995). Estimation of rat body composition by means of electromagnetic scanning is altered by duration of anaesthesia. *Journal of Nutrition*, **125**(6), 1512–20.

Van Loan, M & Mayclin P. (1987). A new TOBEC instrument and procedure for the assessment of body composition: use of Fourier coefficients to predict lean body mass and total body water. *American Journal of Clinical Nutrition*, **45**, 131–7.

Van Loan, M., Belko, A.Z., Mayclin, P.L. & Barbieri, T.F. (1987). *Federation of American Societies for Experimental Biology Proceedings*, **46**, 1334.

Van Loan, M.D. (1990). Biolectrical impedance analysis to determine fat-free mass, total body water and body fat. *Sports Medicine*, **10**, 205–17.

Voltura, M.B. & Wunder, B.A. (1998). Electrical conductivity to predict body composition of mammals and the effect of gastrointestinal contents. *Journal of Mammalogy*, **79**(1), 279–86.

Walsberg, G.E. (1988). Evaluation of a non-destructive method for determining fat stores in small birds and mammals. *Physiological Zoology*, **61**(2), 153–9.

Williams T.D., Monaghan P., Mitchell P.I. *et al.* (1997). Evaluation of a non-destructive method for determining egg composition using total body electrical conductivity (TOBEC) measurements. *Journal of Zoology*, **243**, 611–22.

Witter, M.S. & Goldsmith, A.R. (1997). Social stimulation and regulation of body mass in female starlings. *Animal Behaviour*, **54**, 279–87.

Yanovski, S.Z., Hubbard, V.S., Heymsfield, S.B. & Lukaski, H.C. (eds.) (1996). Bioelectrical impedance analysis in body composition measurement. *American Journal of Clinical Nutrition*, **64** (Suppl.).

Zuercher, G.L., Roby, D. D. & Rexstad, E. A. (1997). Validation of two new total body electrical conductivity (TOBEC) instruments for estimating body composition of live northern red-backed voles *Clethrionomys rutilus*. *Acta Theriologica*, **42**(4), 387–97.

WOUTER D. VAN MARKEN LICHTENBELT

6

The use of bioelectrical impedance analysis (BIA) for estimation of body composition

Introduction

Methods for determination of body composition that are based on the electrical properties of the body include total body electrical conductance (TOBEC: Chapter 5) and bioimpedance analysis (BIA). In this chapter I focus on BIA, more specifically on single frequency BIA (SF-BIA) and multifrequency BIA or bioimpedance spectroscopy (BIS). Once the instrument has been purchased, both SF-BIA and BIS are cheap, and very user and subject friendly. Though the accuracy and usefulness of the methods are still under debate, they are being used increasingly in studies of humans, including sports, clinical work and basic metabolic studies. There have been surprisingly few studies performed in animals.

Bioimpedance analysis belongs to the descriptive methods on a molecular level. Descriptive methods are those that are characterized by a reference method, a well-characterized subject group, and a prediction equation (Wang *et al.*, 1995; Marken Lichtenbelt & Fogelmholm, 1999a). As will become clear in this chapter, BIS has a more functional relation to the water compartments of the subjects than does SF-BIA.

Measurements of electrical conductance in biological tissues date from the end of the nineteenth century (Nyboer, 1959; Baumgartner *et al.*, 1990). The use of bioimpedance for body composition is more recent and was developed by Thomasset in 1962. He used a two-needle electrode configuration at a fixed frequency. Hoffer *et al.* (1969) applied the four-surface electrode method at 50 kHz for measuring TBW. This approach was extended by Nyboer to estimate fat-free mass (FFM = lean mass: LM) and percentage body fat (Nyboer, 1981). Already in the early 1970s it was shown that using impedance measurements at different frequencies could be used to discriminate between extracellular water (ECW) and total body water (TBW) (Ducro *et al.*, 1970; Jenin *et al.*, 1975). These studies have used the impedance measured at a small number of frequencies. From 1992 onwards the first multiple sweep frequency BIA (BIS) studies were published (Cornish *et al.*, 1992a, b; Matthie *et al.*, 1992; Van Loan *et al.*, 1993; Marken Lichtenbelt *et al.*, 1994), following the pioneering work of Kanai *et al.* (1987). This measurement technique is slowly replacing the established SF-BIA.

BIA is based on the principle that the electrical conductance of the body is mainly determined by the water compartments and its solutes. BIA measurements are thus related to body water compartments. Fat mass (FM) and fat-free mass (FFM) are, in most cases, determined assuming a fixed hydration of the FFM. Mean hydration of the FFM in mammals is stable between species (about 0.73) (Wang *et al.*, 1999). However, individual differences can be substantial. This has been observed most frequently in humans (Baarends *et al.*, 1997, 1998; Geerling *et al.*, 1999; Marken Lichtenbelt & Fogelmholm, 1999b), but in other vertebrates, reptiles for instance, individual differences are also large (Marken Lichtenbelt *et al.*, 1993; Drent *et al.*, 1999). In this chapter I will mainly focus on the water compartments. A basic knowledge of electrophysiological theory is required for understanding the principles of SF-BIA and BIS (Schwan, 1957; Ackmann& Seitz, 1984). Here, we have to restrict ourselves to a simplified approach. First, the general theory will be presented, followed by more details with respect to different approaches and calculations used in different studies. Since most present studies concern human body composition, I will first describe the application as it is used in humans. At the end of the chapter, animal studies will be discussed.

Human studies

Single frequency BIA

SF- BIA applies a known alternating current (I) of 500–800 μA and 50 kHz through the body, in most cases from a hand to a foot at the same side of the body (Fig. 6.1). Originally a two-needle approach was used (Thomasset, 1962), followed by the utilization of technically superior four electrode needle or surface electrodes (Hoffer *et al.*, 1969). The voltage (V) is measured from wrist to ankle. Since the current is fixed, using Ohm's Law ($V = IR$), from the voltage (V), the resistance (R) can be calculated.

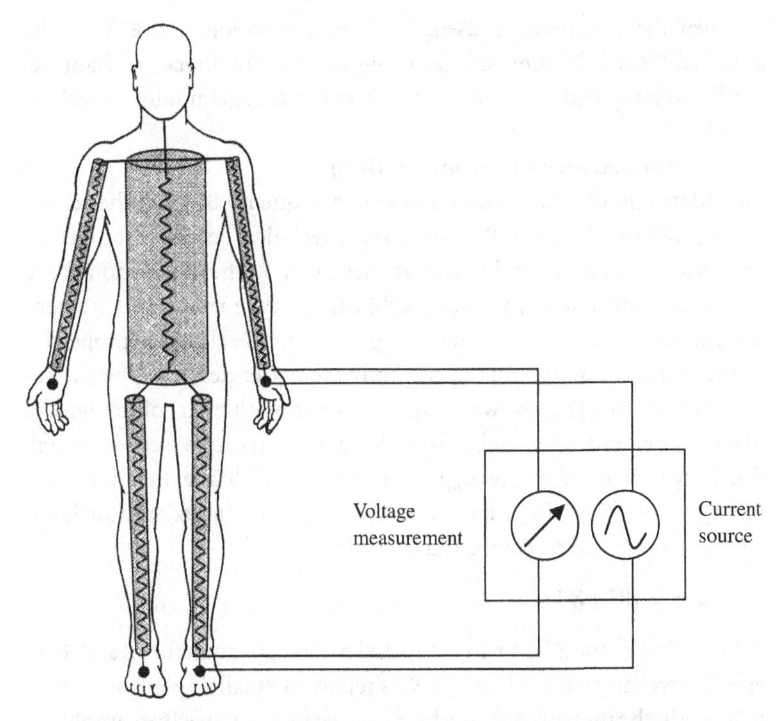

Fig. 6.1. Standard placement of electrodes for whole-body bioimpedance analyses, assuming that the body can be represented by a cylinder. As indicated, the body is more accurately depicted as five cylinders (two arms, trunk and two legs) connected in electrical series. These segments can be measured separately (see under segmental analyses).

Assuming the water compartment of the body is a cylinder, the resistance is related to the length (L) and the radius (r) of the cylinder by:

$$R = \rho L / \pi r^2 \tag{1}$$

where ρ is the specific resistivity of the compartment (mainly body fluid). The volume of a cylinder is given by:

$$\mathrm{Vol} = \pi r^2 L \tag{2}$$

Combining these two formulas results in:

$$\mathrm{Vol} = \rho L^2 / R \tag{3}$$

L^2/R is called the resistivity index and is linearly related to total body water (TBW). Normally in humans L is stature, but sometimes the distance between the electrodes is used. Prediction equations for BIA can be obtained from validation studies using other more direct (mechanistic) methods (Wang *et al.*, 1995; Marken Lichtenbelt & Fogelmholm, 1999a).

Bioimpedance spectrometry (BIS)

To understand the shortcomings of single frequency BIA and the advantages of BIS relative to SF-BIA, some more technical details of the electrical properties of biological tissues are necessary. In the description above the use of Ohm's law was an oversimplification of the actual measurement situation, i.e. the use of an alternating current. The electrical properties of the water compartments in the body are better described by the so-called impedance (Z). The water compartments with their solutes behave like resistors, but the membranes behave like capacitors (see Fig. 6.2(*a*), (*b*)). They have the capacity (C_m) to store electrical charge. Resistors have resistance (R). Capacitors have reactance (X_c). The impedance includes both resistance and reactance and is described by:

$$Z = \sqrt{(R^2 + X_c^2)} \tag{4}$$

In biological tissue X_c is small compared to Z (<4%), therefore Z and R are often interchangeable. Although X_c is relatively small, it plays an important role in the measurement technique used in BIS. Capacitors are able to store an electrical charge for a brief moment. They may be considered as frequency-dependent resistors. Imagine that an alternating current stores electrons on the negative lead of a capacitor. After alternation, the storage of electrons takes place on the other side of the capacitor. The cell membrane charges and discharges at the rate of f (frequency). This reversing of charge takes time. This causes the sinusoidal signal for the voltage to be out of phase with that for the current. The voltage is said to lag behind the current. This phase shift is described as the phase angle ($\Phi = \arctan X_c/R$). It can be understood now that, at direct current and low frequencies, the membranes block the current (after the capacitor is charged no current flows). Only the extracelllar compartment conducts the current. When the frequency increases, e.g. from 1 kHz to 100 MHz the capacitor effect of the membrane functions and reactance gradually increases and at continuing increasing frequency, decreases again (Fig. 6.3). At very high frequencies the current alternates so fast that the storage of charge is negligible and the current just passes the membrane without a voltage phase shift. Both extracelluar (ECW) and intracellular water (ICW) conduct the current. The reactance is low again. Thus, at both

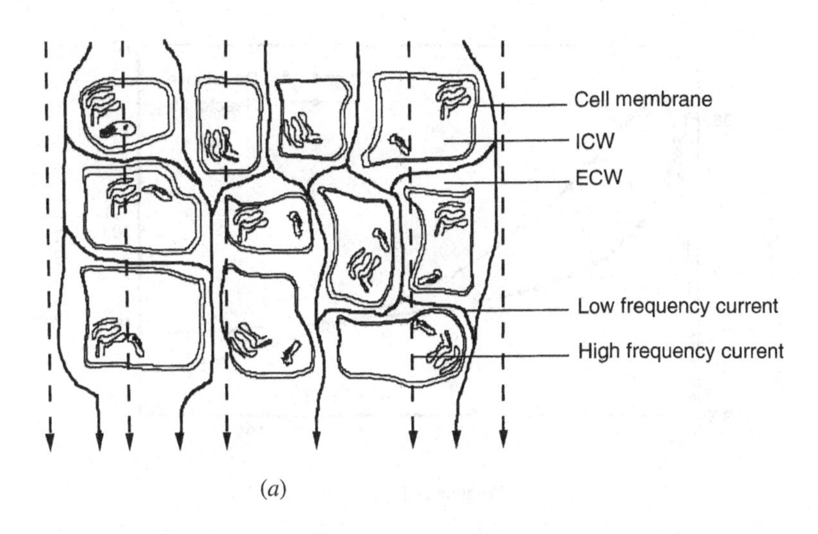

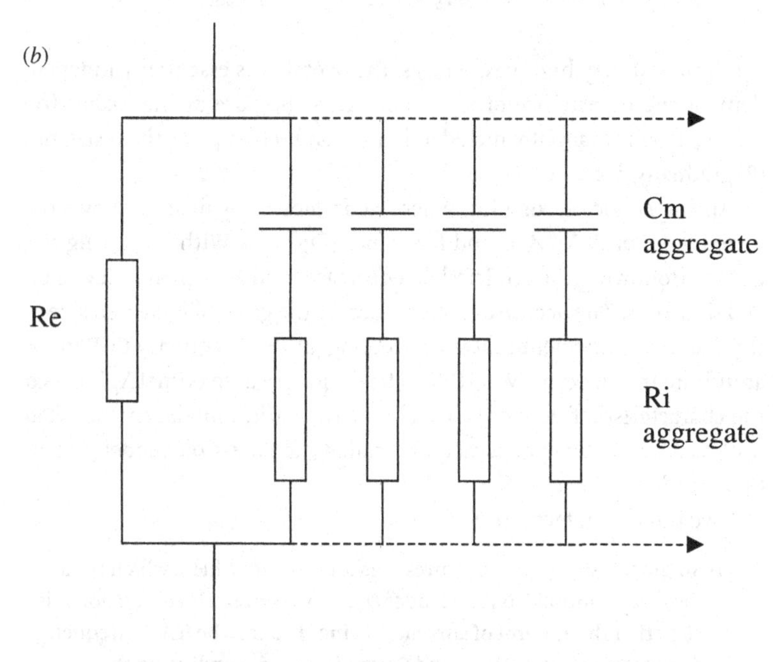

Fig. 6.2. (*a*) Diagram representing high-frequency and low-frequency current distribution in a cell suspension. (*b*) A circuit-equivalent model, analogous to the Cole–Cole model, for conduction of an alternating electric current in a cell suspension. Recw (Re in figure) is the component value of the extra cellular fluids; Ricw (Ri in figure) is the aggregate component value of all cells in suspension of the intracellular volumes (C_m: capacity cell membrane).

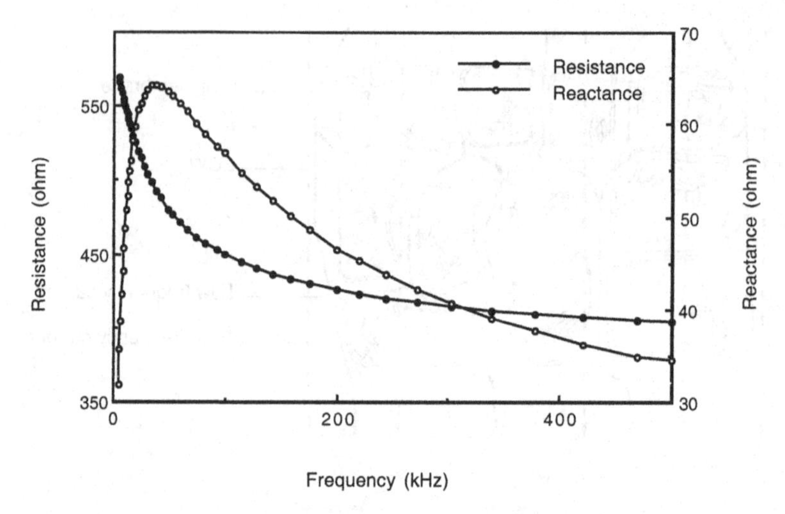

Fig. 6.3. Whole body resistance (R) and reactance (X_c) plotted against the frequency of the current from 5–500 kHz of an adult male.

very low and very high frequencies, the overall *Z* is essentially independent of the capacitance of the membranes. Because at increasing frequency ICW is gradually included in the conduction path, the resistance (*R*) gradually decreases (Fig. 6.3).

An impedance locus plot, where *Xc* is plotted against *R*, shows the dependence of *R*, X_c, *Z*, Φ and frequency (Fig. 6.4). With increasing frequency (following the semicircle), *R* decreases and X_c increases, levels off and decreases. The decrease of *R* is caused by the gradual increased size of the water compartment that is conducting (at zero frequency ECW only, at infinite frequency ECW + ICW). The frequency at maximal X_c is called the characteristic frequency (ω_c). The mathematical model that is often used to describe this phenomenon is called the Cole–Cole model (Cole & Cole, 1941).[1]

Two important facts are evident.

(i) Single frequency BIA measures in most cases at 50 kHz, which is the mean ω_c of muscle tissue (Geddes & Baker, 1967; Settle *et al.*, 1980). It is clear that the amount of current flowing through the ICW is frequency dependent. Changes in ω_c and C_m may occur, depending on the hydration status, the ECW/ICW ratio of the subject and also

[1] Cole–Cole model (Fig. 6.4): $Z = R + (Ro - R_\infty)/[1 + (j\omega/\omega_c)^\alpha$, where j stands for the imaginary number and α is a dimensionless numerical constant; $Recw = Ro$ and $Ricw = 1/(1/R_\infty - 1/Ro)$.

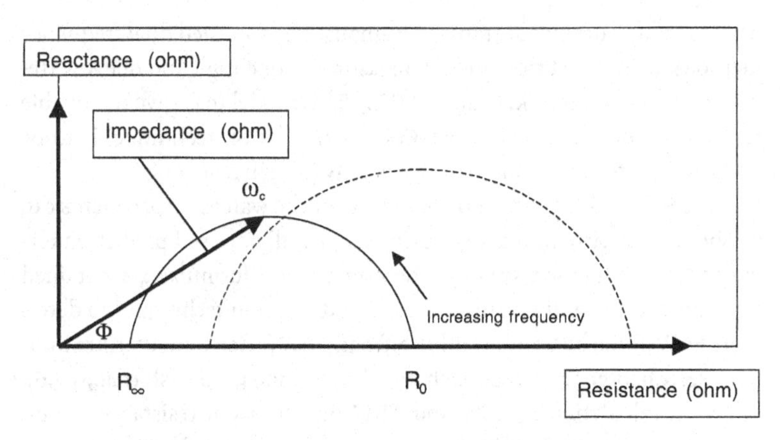

Fig. 6.4. Reactance plotted against resistance with increasing current frequency (the small semicircle). Indicated are the reactance (Xc), resistance (R), impedance (Z), phase angle (Φ), and characteristic frequency (ω_c). $R_{ecw} = R_0$; $R_{tbw} = R_\infty$; $R_{icw} = 1/(1/R_\infty - 1/R_0)$. The large semicircle indicates what would happen when ECW is increased or when ionic concentration is decreased.

interindividual differences are known to exist (Fig. 6.4: curve shift) (DeLorenzo *et al.*, 1997; Earthman *et al.*, 2000). For instance ω_c of muscles may vary widely between individuals from 30 to more than 100 kHz (Geddes & Baker, 1967; Cornish *et al.*, 1993). Consequently the contribution of ICW to the measurement in SF-BIA is variable.

(ii) The Cole–Cole plot shows two intersections with the X-axis (Fig. 6.4). One at zero frequency, revealing R_0 determined by ECW (Recw), and one at infinite frequencies revealing R_∞ determined by TBW, i.e. by both ECW and ICW (Rtbw).

Because of technical reasons, measurements at zero and low frequency and at very high frequencies should be avoided. BIS makes use of a series of measurements that are logarithmically spaced from 1–5 kHz to about 500 kHz. Several studies use these measurements to model according to Cole–Cole, where resistance values at zero and infinite frequency are obtained by extrapolation (Matthie *et al.*, 1992; Marken Lichtenbelt *et al.*, 1994; Van Loan *et al.*, 1995; DeLorenzo *et al.*, 1997). Others believe that the impedance at ω_c gives a better TBW estimate (Cornish *et al.*, 1992a, b, 1993). These opinions and arguments are described by Stroud *et al.* (1995) and Matthie & Withers (1996).

In both cases one reveals theoretically based estimates of R_{ecw}, R_{tbw}, and derived from those: R_{icw}. This makes it possible to study, not only TBW and body fat, but also water distribution. ECW/ICW ratio is often used to study nutritional status in humans. The so-called dual frequency methods, using the Fricke model, measures at one high and one low frequency (for instance, 5 kHz and 1 MHz). This model may give reasonable results, but BIS (using the Cole–Cole extrapolation technique) is to be preferred especially for the estimate of ICW (Jaffrin *et al.*, 1997).

Once R_{ecw} and R_{tbw} are known, the size of the water compartments can be obtained by two means: first, using empirically derived prediction formulas from validation studies. These prediction formulas are obtained from validation studies using both BIS and dilution techniques to determine ECW (often by bromide dilution) and TBW (often by deuterium dilution: see Chapter 3; Marken Lichtenbelt *et al.*, 1994; Cornish *et al.*, 1996). Secondly, calculation of ECW and TBW directly from resistance values, using specific resistivities of the water compartments and taking mixture effects into account, i.e. the mixture of the conductor and the non-conductor (Hanai, 1968). These mixture effects are greater at low frequency, since ECW (the conductor) represents only 25% of the total body volume (75% non-conductor), while at high frequency TBW (the conductor) is about 60%. For instance, at 1 kHz (low frequency) plasma (ECW) is 4–6× more conductive than ECW of skeletal muscle. The specific resistivity of ECW (p_{ecw}) of skeletal muscle is increased, because the cells are non-conductive at LF and restrict the current of the ECW surrounding the cells. Taking this theory (Hanai, 1968) into consideration, equations have been constructed for both ECW at low frequency and ECW+ICW at high frequencies (Van Loan *et al.*, 1993; DeLorenzo *et al.*, 1997). This procedure replaces the empirical determined population-specific equations. However, the specific resistivities (p) of ECW and ICW may be population specific, because of differences in the (ionic) composition of the water compartments. Secondly, body shape (relative size of the different segments of the body) can be population specific. The contribution of the specific resistivities and the different segments of the body to the impedance measurement will now be explained.

The equations to calculate the water compartments, coupled with the mixture theory (Hanai, 1968) are for ECW:

$$ECW = k_{ecw}(H^2.Wt^{0.5}/R_{ecw})^{2/3} \qquad (5)$$

H is height in cm, Wt is weight in kg and the $k$ecw (in Ω.cm) is defined as

$$k_{ecw} = 10^{-3}.(K_b{}^2.p_{ecw}{}^2/D_b)^{1/3} \qquad (6)$$

K_b is a body geometry factor that relates the relative volumes of the legs, arms, and trunk, p_{ecw} (in Ω.cm) is the resistivity of the ECW and D_b (in kg/m³) is the total body density. Assumed values, which have been scaled with deuterium dilution and bromide dilution (Van Loan *et al.*, 1995), are: $K_b = 4.3$; $D_b = 1.05$; $p_{ecw} = 214$ (males), 206 (females), resulting in k_{ecws} of 0.306 (m) and 0.316 (f).

The equation for calculation of ICW is:

$$(1 + V_{icw}/V_{ecw})^{5/2} = [(R_e + R_i)/R_i].[1 + k_p V_{icw}/V_{ecw}] \quad (7)$$

kp is the resistivity ratio defined as p_{icw}/p_{ecw}.

Assumed values are (Van Loan *et al.*, 1995): p_{icw} is 824 (m) and 797 (f), resulting in k_p of 3.82 (m) and 3.40 (f).

Agreement between BIS and dilution methods for adult humans has been best achieved when tissue resistivity values have been recalculated for each specific population. This results in different values of *kecw*, and *kp* in different studies (Smye *et al.*, 1994; Bedogni *et al.*, 1996; Armstrong *et al.*, 1997; DeLorenzo *et al.*, 1997; Janssen *et al.*, 1997; Marken Lichtenbelt *et al.*, 1997; Ellis & Wong, 1998). The largest validation study carried out to date is by Ellis & Wong (1998), comparing BIS with dilution techniques for a large population of children, adolescents, young adults of both genders and of varying ethnicity. They conclude that BIS can be recalibrated on a group-to-group basis to achieve approximately zero mean differences between ECW, ICW and TBW estimates when compared with dilution based estimates. This however, does not necessarily ensure that the BIS estimates are accurate for any subsequent studies in a different population or for individuals of the same population. For BIS to be universally applicable, the basic assumptions should be independent of the population of which they were derived. An important parameter in this respect is Kb, being a body geometry factor that relates the relative volumes of the leg, arms and trunk. The assumptions underlying the estimate of this parameter can be improved making segmental analyses.

Segmental measurements

Most measurements take place from wrist to ankle (distal electrode placement), assuming that the body is a cylinder. However, the body is not at all a perfect cylinder, but is better represented as five cylinders (two arms, two legs and trunk) with large differences in resistance (Fig. 6.1). Arm, trunk and leg are connected in series. The arm, which is the conductor with the smallest cross-sectional area, determines most of the resistance of the series. The arm is about 4% and the leg about 17% of body weight, while they account for about 47% and 50% of the whole body resistance when measured from wrist to ankle. The trunk has hardly any influence on the whole body resistance, but comprises 46% of the body weight (Patterson *et al.*, 1988; Baumgartner *et al.*, 1989). Therefore, studies have been carried out using electrodes placed more proximally (Scheltinga *et al.*, 1991). Correlation with TBW was substantially improved with proximal electrode placement (Marken Lichtenbelt *et al.*, 1994). A drawback of this procedure is that some parts of the body are not included (forearm and lower leg). In this respect it is interesting to note that whole body measurements reveal only marginally better correlations between the resistivity index and TBW, than using (arm length)^2/Rarm only (Settle *et al.*, 1980; Baumgartner *et al.*, 1989)! These authors conclude that the prediction of TBW using the whole body BIA is dependent on their strong association with the mass and bioelectric characteristics of the appendicular skeletal muscle.

It therefore seems to make sense to measure in separate body segments, combined with appropriate weighing factors, to calculate total conducting volume (Davies *et al.*, 1988; Smith, 1993). Especially so since a physiologically significant complication of whole body BIA or (one) segmental BIA is that the distribution of ECW in the body can be affected by the position of the body. Although the distribution between ICW and ECW is known to be unaffected by changes in body position (Maw *et al.*, 1995), there are considerable changes in the regional distribution of ECW (Rowell, 1993; Lundvall *et al.*, 1996), so that the ECW is heterogeneous with respect to local tissue resistance. Mean differences between ECW values, determined by whole body impedance, can be more than 7.4 ± 1.6 *L* depending on standing and supine phases. The effect of changes in body position, and/or individual differences in size of the segments, can be overcome by calculating ECW from BIS measurements of five segments separately, each with their own specific resistivity (Zhu *et al.*, 1998). A practical six-electrode technique for segmental BIS has been described (Organ *et al.*, 1994).

Instead of using eqns. 5 and 6 (see above) Zhu *et al.* (1998) used for each segment a different formula and summed these volumes. Moreover, they incorporated estimates of body water volumes of those parts of the body that are not measured by BIS (head, neck, hands, feet). Using this model, they were able to show that ECW (sum of segments) was almost independent of changes in body position. It is also a prerequisite for applications where continuous fluid monitoring is wanted, e.g. hemodialysis. There are, however, indications from other studies that even segmental analyses may give rise to errors, mainly because the electrical field in the trunk is inhomogeneous (Kreel *et al.*, 1998). Nevertheless, the segmental technique seems to make BIS more generally tenable and less population (species) specific than whole body measurements.

Measurement conditions

Measurement conditions are very important because variations have been shown to introduce discrepancies. The control of the BIA measurement conditions assures an accurate, reproducible and valid interpretation of the results. Factors known to introduce physiological meaningful variability in impedance measurements include type of instruments, body position, recumbence time, electrode placement, body temperature, ambient temperature (skin temperature) and exercise.

Bioimpedance analysers

Each BIA instrument houses an alternating current source, cables and electrodes for applying the current to the body and for sensing the voltage difference (and depending on the instrument other variables like phase shift). It was only around 1985 that the first single frequency commercial instrument by RJL, Inc. (Detroit, MI) was available. Gradually other manufactures introduced BIA analysers, most of which are single frequency devices, some use two or three different frequencies. It was in 1991 that the first commercial multifrequency instrument by Xitron Technologies was introduced (see Table 6.1).

Another distinction is between the two-electrode bridge technique and the four skin-electrode configuration. The two-electrode bridge technique combines the current and voltage electrodes. In this configuration measurements can be done at very low currents. The disadvantages are (i) electrode polarization, that may influence the measured impedance and (ii) the need for needle electrodes because of the high resistance of the skin. Needle electrodes obviously are more invasive then surface electrodes. They may cause trauma especially when repeated measurements are applied. In animal studies needle electrodes can be useful, when surface electrodes cannot be used due to fur or feathers.

The four surface electrodes method is non-invasive. The objective of this configuration is to measure Z independently of the surface impedance. This is achieved by introducing the test current through one set of electrodes and by measuring the voltage at a second set, placed a few centimetres more proximal from the corresponding current electrode.

Table 6.1. *Bioimpedance analysers, general characteristics and suppliers*

Single frequency analysers (whole body)
BODYSTAT 1500; Bodystat Ltd, UK
Biodynamics 310e/550; Biodynamics Corporation, USA
E-Z Comp 1500; Cranlea and Co, UK
Maltron BT-905; Maltron Ltd, UK
RJL BIA 101Q/101AS/spectrum II; RJL Systems Inc., USA
Valhall 1990B; Valhall Scientific, USA
SEAC BIM 4; UniQuest Limited, Aus
AKERN 101; Akern s.r.i., USA
Single frequency analysers (arm to arm or foot to foot; for human use only)
Tanita TBF 215/300/310/410 GS; Tanita Corporation, USA
Omron HBF-300; Omron Healthcare Inc., USA
Yamato DF-311; Yamato Scale Co., Ltd, Japan
Dual (triple) frequency analysers
HUMAN IM SCAN; Dietosystems, Italy
BODYSTAT MultiScan 5000; Bodystat Ltd, UK
BMR 2000; BMR Co., USA
Multifrequency analysers (BIS)
SEAC SFB3; Uniquest Limited,Aus
Xitron 4200; Xitron Technologies Corp., USA

Note:
This list is not intended to be exhaustive

Several studies report systematic differences in resistance readings on the same subjects with different apparatus, e.g. Deurenberg *et al.* (1989), and Heitmann (1990). In a study that compared six analysers, using an electronic circuit that simulated human body impedance, analyser errors varied from <1% to nearly 20% (Oldham, 1996). Caution should also be exercised when comparing different instruments using electronic circuits or components, that are supplied with the instruments. These electronic circuits often do not mimic the actual measurement situation, and thus errors may only become apparent when a validation is performed on live subjects (Heitmann, 1994).

Body temperature

BIA is strongly temperature dependent. The impedance is negatively related to temperature. Thus recent exercise should be avoided, since during, and shortly after, exercise body temperature will be increased and this will decrease resistance (Khaled *et al.*, 1988). Environmental temperature fluctuations should be minimized, because this affects the skin temperature, which in turn affects the BIA result (Caton *et al.*, 1988; Gudivaka *et al.*, 1996). Heating the skin, using single frequeny BIA increased predicted TBW by 2.6 l compared to baseline measurements (Gudivaka *et al.*, 1996). The authors reason that the main effect of skin temperature can be explained by a change in conductance of the skin, and to a lesser extent to compartment fluid shifts. They conclude that an acceptable range in temperature is ±2.7 °C, limiting the variation of TBW to ±1%. In general, it is advised to perform BIA measurements at a standardized time (Rodríguez *et al.*, 1999). In the morning, after an overnight fast (see below) is advised. This may reduce minor effects of diurnal temperature fluctuations and diet-induced thermogenesis.

Recent exercise

Measurements must be carried out in standardized fashion, preferably while the subject is laying supine; the person must be still and breathing normally. During and shortly after exercise no accurate measurements of BIA or BIS can be done, because this would affect BIA by at least three hypothesized mechanisms (Kushner *et al.*, 1996):

(i) The hemodynamic response to exercise consists of increased cardiac output and blood flow to skeletal muscles. The increased vascular perfusion will reduce Z (this problem can be overcome by segmental measurements). Increased muscle temperature will also cause a decrease in Z.

(ii) Heat dissipation includes increased cutaneous blood flow by vasodilatation. Again, this will result in a decrease in Z. The rise of skin temperature and sweat production will reduce Z.

(iii) Sensible and insensible fluid losses result in dehydration, loss of TBW, and an increase in Z. Depending on the goal of the study, dehydration should be avoided.

In an experiment using single frequency BIA, Liang and Norris (1993) examined the effect of exercise without significant changes in hydration status. They found that exercise decreased *R* by 3%.

Food and beverage

Measurements shortly after a meal and or drink potentially can influence the results. However, it has been shown that a surplus of 2 l of fluid in the abdominal cavity has little or no effect on Z by whole body BIA, because of the large cross-sectional area of the trunk and the relatively small contribution of the trunk to whole body Z. However, using segmental measurements, the influence of additional fluids due to drinks are expected to be larger. In general, studies on consumption of food and beverages are conflicting, showing a small but significant effect (Deurenberg *et al.*, 1988; Gallagher *et al.*, 1998), or no effect at all (Steen *et al.*, 1987; Eilsen *et al.*, 1993). When BIA is used to calculate FM, the latter may be overestimated because of the weight increase. The effects of food can be omitted by measuring after a fast of 8 h or more (preferably an overnight fast). For related reasons, it is advised to empty the bladder prior to measurement.

Body position

Body position, e.g. supine vs. standing may affect body water distribution, which can partly be overcome by segmental analysis. Measurements are almost invariably made with the subject lying down. The time spent lying supine influences whole body BIA. Impedance values rise sharply within the first 10 minutes after the subject assumes the supine position and then continue to rise more gradually for up to 4 hours. Therefore, standardization with respect to timing of measurement after recumbence is necessary. In practice, however, many different intervals are applied. To prevent shortcut of the electrical circuit, both arms and legs should be abducted at about 30 degrees angle from the trunk axis.

Electrode placement

Placement of electrodes should be at defined sites and the procedure carefully practised to increase observer reliability (see, for instance, Chumlea *et al.* 1994). The present convention is to place two distal current-introducing electrodes on the dorsal surfaces of the hand and foot proximal to the metacarpal–phalangeal and metatarsal–phalangeal joints, respectively. In addition, two voltage sensing electrodes are applied at the pisiform prominence of the wrist and between the medial and lateral malleoni of the ankle (Anon, 1996). Segmental analyses include other sites. Surface electrodes should be of good quality. Spot or foil electrodes, similar to those used for electrocardiography can be used, but must have a minimum surface area of 4 cm^2 for adults. Some manufacturers advise using elongated band electrodes (length approximately 4–5 cm) in order to obtain a more homogeneous electrical field. In young children and prematures, the size and the position of the electrodes deviate, because of anatomical reasons.The skin needs to be prepared with alcohol before placement of electrodes.

Shape of the body and disease

The most established prediction equations for BIA are based on large heterogeneous groups. These equations are not able to take into account special subgroups as, for instance, the obese (Gray *et al.*, 1989; Rising *et al.*, 1991), children (Houtkooper *et al.*, 1992) or certain patient groups (Bedoggni *et al.*, 1996a, b; Baarends *et al.*, 1997, 1998). In order to establish more accurate results, many authors come up with special equations for the subgroup of interest: children (Houtkooper *et al.*, 1992) , sex-specific equations (Kushner & Schoeller, 1986), low birthweight infants (Mayfield *et al.*, 1991; Raghavan *et al.*, 1998), Pima Indians (Rising *et al.*, 1991), malnutrition (Pencharz & Azcue, 1996), growth hormone-deficient adults (Binnerts *et al.*, 1992). One should be warned against the use of these specific equations, because in case of changes after treatment, the patients may move from one prediction equation (of patients) to the other (of healthy subjects) (Rising *et al.*, 1991).

In conclusion there are several reasons why BIA can be population specific:

SF-BIA

As discussed above, population specificity in SF-BIA can be attributed to individual differences in ω_c. BIS is theoretically superior with this respect.

Body water distribution

Since the body is not a perfect cylinder, differences in body water distribution will lead to errors in case of whole body BIA. Better estimates can be obtained by using segmental measurements. However, as described above, the trunk measurements still pose problems, because of the inhomogeneous fluid distribution.

Composition of the water compartments

Alterations of serum electrolytes, and hematogrid may change specific resistivity independent of fluid volume. A 5 mmol change in ion can affect the predicted ECW 1–2%, and ICW 4–5% (Scharfetter *et al.*, 1995). Since ionic effects on resistivity are linear, they could be accounted for.

Direction of fibre

The resistivity of muscle measured transverse to the direction of the fibre is several times higher than that measured along the direction of the fibre (Geddes & Baker, 1967). This may also explain the relatively high resistivity in the trunk compared to the limbs (Chumlea *et al.*, 1988).

Validation and accuracy

Single frequency BIA

In humans, SF-BIA should be validated against TBW by deuterium dilution, for derivation of a prediction equation. A prediction equation for adults males, for example, may appear as follows: TBW $= 0.63\ (L^2/R) + 2.03$, with an error of estimate (SEE): 2.03 *L* (Lukaski *et al.*, 1985). If we measure a male (weight 90 kg, stature 174 cm), and *R* is 505 ohm at 50 kHz, TBW according to the Lukaski equation amounts to: TBW $= 0.63\ (174 \times 174) + 2.03 = 39.8$ *L*. BIA is, however, often validated against densitometry or multicompartment models, relating L^2/R (or L^2 and $1/R$ separately) to FFM. Often additional independent variables are included, like age, BM and/or sex in the prediction equation. For instance, in males: FFM $= 0.0013\ L^2 - 0.044\ R + 0.305$ BM $- 0.168$ age $+ 22.668$, SEE $= 3.61$ kg (Segal *et al.*, 1988). The addition of these factors may adjust for anthropometric differences between individuals, e.g. relative trunk size or different composition of water compartments. In these cases, the relation between the resistance measurement and FFM becomes obscured by the many extra parameters involved, and becomes less generally tenable. This kind of prediction equation is strongly population dependent.

The accuracy of a prediction equation is determined when the equation is applied to independent samples (preferably from another laboratory) in cross-validation studies, which was the case in the equations presented above. At the moment, there are more than 30 published different prediction equations available for humans (Heitmann, 1994; Baumgartner, 1996; Houtkooper *et al.*, 1996). When scientifically based, validation studies should primarily be focused on TBW. However, there are many prediction equations for FFM, percentage body fat, ICW, and body cell mass. Correlation coefficients are nearly always significant ($P < 0.05$), and often high ($r > 0.95$). The relationships reported may have simply been a result of a high intercorrelation between the variables (TBW, ICW and body cell mass).

In general, regression analysis does not provide much information about accuracy on a group or individual level. A good way to present the data is a so-called Altman and Bland analysis, comparing differences between alternative methods with the average values obtained by the two methods (Altman & Bland, 1983). This provides both the bias (or off set value = average difference between measured and predicted) and the error (= SQRT (SUM(predicted − measured)2/*n*)). An alternative for the error value (or accuracy on individual level) is the standard error of estimate (SEE). In the validation studies mentioned above, criterion method (or gold standard) for TBW was deuterium dilution, and for FFM was densitometry by under water weighing.

General prediction equations, based on cross-validation selected by Houtkooper (Houtkooper *et al.*, 1996), reveal for TBW an error ranging from 0.23–2 l with a bias of 0.05 to −1.00 l. For FFM, these numbers are: error: 1.7–3.3 with a bias ranging from −1.7 to 1.4 kg. Assuming a male subject with 20% body fat and a hydration of FFM of 73%, the error for TBW thus approaches 4.5%, and for the FFM 8%. The effect of the bias can have an additive effect. From these errors it can be concluded that the method is relatively inaccurate for individual measurements. Deuterium dilution has an error of 1% for TBW and of approximately 2% for FFM (Westerterp *et al.*, 1995; Schoeller, 1996). This means that for actual TBW the BIA error reaches 5.5%.

Bioimpedance spectrometry

Although BIS is better equipped than SF-BIA and is theoretically superior, assumptions are necessary regarding specific resistivity of the conductive material. This may have accounted for the fact that, until now, not much improvement with respect to the accuracy has been made. With proper electrode placement and good measurement conditions (see below) no large errors are expected to occur in obtaining R_{ecw} and R_{tbw}. The extrapolation technique using the Cole–Cole model reveals a precision of <2% for R_{ecw} and R_{icw} estimates for an individual. For ECW, SEE values range from 0.6 l (Marken Lichtenbelt *et al.*, 1994) to 2.3 l (Ellis & Wong, 1998), for TBW from 1.5 to 2.7 l, and for ICW from 1.1–2.5 l. The bias found by Ellis, using default specific resistivities, was for ECW 0.3 l (males) and −0.6 l (females), TBW −2.7 l (m) and −2.4 l (f), ICW −2.8 l (m) and −1.8 l (f). As discussed above, with respect to the bias, scaling to reference values is possible by adjusting the specific resistivities of ECW and ICW. However, this procedure makes BIS population specific. Segmental anaysis combined with BIS may be an improvement, but validation studies have not yet been carried out.

Fluid changes

Measurement of changes in body composition due to diet, treatment or environmental factors may be relatively accurate (de Vries *et al.*, 1987; Van Loan *et al.*, 1995; Jaffrin *et al.*, 1997; Earthman *et al.*, 2000). This is due to the fact that assumptions linked to the method before and after intervention are largely comparable. As explained above (effect of fluid shift on membrane capacitance), BIS seems superior to SF-BIA in this respect. This is confirmed in the few validation studies carried out so far (Van Loan *et al.*, 1995; Earthman *et al.*, 2000).

Animal studies

Both SF-BIA and BIS have hardly been applied in wild animals. Most studies with SF-BIA are with farm animals and the investigators concluded, on the basis of correlations with carcass analysis, that BIA is able to predict skeletal muscle mass in cattle (Marcello *et al.*, 1994), pigs (Marcello *et al.*, 1992), and sheep (Cosgrove *et al.*, 1988). Recent studies on whole body SF-BIA in wild animals have been performed in seals (Pietraszek & Atkinson, 1994), bears (Farley & Robbins 1994; Hildebrand *et al.*, 1998), and the herbivorous southern hairy-nosed wombat (Woolnough *et al.*, 1997). In the study of seals the technique was used to study the potential of the method to monitor estrogen levels by detecting estrogen mediated hydration of reproductive tissues. Hildebrand *et al.* (1998), and Farley and Robbins (1994) both used SF-BIA with needle electrodes, and deuterium dilution with carcass analyses as the reference method in wild bears. The average BIA and dilution values were close to carcass analysis values; but Hildebrand (Hildebrand *et al.*, 1998) concluded that dilution was better because of a lower SEE. Though some studies show good correlations between SF-BIA and dilution or carcass techniques, to date, no well-performed cross-validation studies with animals have been carried out.

The use of BIS in animals is restricted to rats (Cornish *et al.*, 1992a, b; Cha *et al.*, 1995) and cattle (Thomson *et al.*, 1997). In anaesthetized rats, using tetrapolar electrode configuration with needle electrodes, whole body Z and Φ were measured at six different frequencies. L^2/Z_c and L^2/Z_0 were correlated with TBW and ECW by dilution techniques and carcass analyses, respectively. Correlations were high with SEE of 6.5% and 3.2% for TBW and ECW respectively (Cornish *et al.*, 1992a, b). Another study in rats indicated that changes in the ECW by infusion were detected in the legs, but not in the trunk (Cha *et al.*, 1995). The authors ascribe their results to compartmentization and to segmental differences in the orientation of muscle and organ fibres. They did not discuss the use of general mixture theory.

In an experiment in healthy cattle, SF-BIA correlated well with body water compartments by carcass analyses, but not so in a group of underfed cattle (Thomson *et al.*, 1997). In this study, however, whole body measurements were applied from shoulder blade to pin bone. Segmental measurements most probably would improve the measurements and would make the measurements more general and thus also suitable for underfed or diseased cattle. More experiments are needed.

In general, in animal studies special attention is needed to address the following:

Electrode placement

As shown in human studies, proper placement of electrodes is important. Since not many studies in animals have been carried out, pilot experiments for determination of the optimal location of electrodes need to be performed. Electrodes in humans should be at least 5 cm apart. In many animal species it will be more practical to use the needle electrodes inserted in the skin.

Length to cross-section area

Of the segments, the ratio of length to cross-section area must preferably be large. Snakes seem to be the perfect study animals (though body temperature should be controlled).

Subcutaneous fat

Some studies indicate that a large insulating subcutaneous fat layer may pose problems.

Body temperature

Body temperature in animals (mammals, birds and reptiles) is subject to variations depending on season, time of the day, environmental factors and exercise. As indicated above, body temperature should be standardized.

Skin insulation

Depending on the type of skin, the resistance of the skin may have a large impact on the measurements. Needle electrodes may be necessary.

Hydration FFM

If body water values are to be converted to FFM, specific hydration of FFM must be taken into account.

Gut fill

Preferably BIA is applied in a fasted condition. In animals, the gut fill can be substantial and it may take some time before a fasted state is reached.

Quietness

The animal must be still, quiet and breathing normally during measurement. In practise many animals will be stressed. With reptiles it may be possible to measure them at standardized low body temperature, this will ensure they are calm. Other animals must be anaesthetized. However, there are indications that BIA may be affected by the depth of the anaesthesia (Farley & Robbins, 1994), partly due to a drop in body temperature.

Conclusions

Most BIA studies have concerned human body composition, therefore this chapter has focused mostly on the application of the method as it is used in humans. At the end of the chapter animal studies were presented and suggestions for animal studies are given. BIA is based on the principle that the electrical conductance of the body is mainly determined by the water compartments and solutes. I have distinguished single frequency (SF-BIA) and multifrequency BIA (or bioimpedance spectrometry BIS). SF-BIA applies a known alternating current through the body, and measures the resistance. The $length^2$ divided by the resistance is related to total body water. Empirically derived prediction equations are used. Validation studies, using deuterium dilution or densitometry as reference methods), indicate that SF-BIA is not accurate on an individual level, but can be used in epidemiological studies. SF-BIA is strongly population specific.

Bioimpedance spectrometry (BIS) is a multifrequency method, making use of the fact that cell membranes behave like capacitors that block direct current and let the current pass at high frequencies. It applies a series of measurements at different frequencies. Using filter characteristics, it is thus possible to discriminate between the resistance of ECW (extrapolation of the measurements to direct current) and the resistance of TBW (extrapolation of the measurements to current of an infinite frequency). L^2/R_{ecw} can be validated against bromide dilution and L^2/R_{tbw} against deuterium dilution. Although BIS holds promise compared to single frequency, studies so far do not show much improvement with respect to accuracy. When the specific resistivity of the water compartments is known, it appears to be possible to calculate the water compartments directly from the R values. It needs to be investigated whether the general calculation can be applied to different populations, since the specific resistivity may vary according to the population under study. Segmental analyses seems to improve the accuracy of BIS. Finally, BIS may be useful measuring body composition changes. BIA (and BIS) measurements should be standardized regarding temperature, electrode positioning, exercise, eating and drinking. Once the instrument is available, BIA is cheap, and very user and subject friendly.

REFERENCES

Ackmann, J. J. & Seitz, M. A. (1984). Methods of complex impedance measurements in biologic tissue. *Critical Reviews in Biomedical Engineering*, **11**, 281–311.

Altman, D. G. & Bland, J. M. (1983). Measurement in medicine: the analysis of method comparison studies. *The Statistician*, **32**, 307–17.

Anonymus (1996). Bioelectrical impedance analysis in body composition measurement: National Institutes of Health Technology Assessment Conference Statement. *American Journal of Clinical Nutrition*, **64**(S), 524–32.

Armstrong, L. E., Kenefick, R. W., Castellani, J. W. *et al.* (1997). Bioimpedance spectroscopy technique: intra-, extracellular, and total body water. *Medicine and Science in Sports and Exercise*, **29**, 1657–63.

Baarends, E. M., Schols, A. M. W. J., Marken Lichtenbelt, W. D. v. & Wouters, E. F. M. (1997). Analysis of body water compartments in relation to tissue depletion in clinically stable patients with chronic obstructive pulmonary disease. *American Journal of Clinical Nutrition*, **65**, 88–94.

Baarends, E. M., Marken Lichtenbelt, W. D v., Wouters, E. F. M. & Schols, A. M. W. J. (1998). Body water compartments measured by bioelectrical impedance spectrometry in patients with chronic obstructive pulmonary disease. *Clinical Nutrition*, **17**, 15–22.

Baumgartner, R. N. (1996). Electrical impedance and total body electrical conductivity. *Human Body Composition*, pp. 79–107. Champaign: Human Kinetics.

Baumgartner, R. N., Chumlea, W. C. & Roche, A. F. (1989). Estimation of body composition from bioelectric impedance of body segments. *American Journal of Clinical Nutrition*, **50**, 221–6.

Baumgartner, R. N., Chumlea, W. C. & Roche, A. F. (1990). Bioelectric impedance for body composition. *Exercise Sport Science Review*, **18**, 193–224.

Bedoggni, G., Merlini, L., Ballestrazzi, A., Severi, S. & Battistini, N. (1996a). Multifrequency bioelectric impedance measurements for predicting body water compartments in duchenne muscular dystrophy. *Neuromuscular Disorders*, **6**, 55–60.

Bedoggni, G., Polito, C., Severi, S. *et al.* (1996b). Altered body water distribution in subjects with juvenile rheumatoid arthritis and its effect on the measurement of water compartments from bioelectrical impedance. *European Journal of Clinical Nutrition*, **50**, 335–9.

Binnerts, A., Deurenberg, P., Swart, G. R., Wilson, J. H. & Lamberts, S. W. (1992). Body composition in growth hormone-deficient adults. *American Journal of Clinical Nutrition*, **55**, 918–23.

Caton, J. R., Molé, P. A., Adams, W. C. & Heustis, D. S. (1988). Body composition by electrical impedance: effect of skin temperature. *Medical Science Sports Exercise*, **20**, 489–91.

Cha, K., Hill, A.G., Rounds, J.D. & Wilmore, D.W. (1995). Multifrequency bioelectrical impedance fails to quantify squestration of abdominal fluid. *Journal of Applied Physiology*, **78**, 736–9.

Chumlea, W.C., Baumgardner, R.N. & Roche, A.F. (1988). Specific resistivity used to estimate fat-free mass segmental body measures of bioelectrical impedance. *American Journal of Clinical Nutrition*, **48**, 7–15.

Chumlea, W.C., Guo, S.S. *et al.* (1994). Reliability for multifrequency bioelectric impedance. *American Journal of Human Biology*, **6**, 195–202.

Cole, K.S. & Cole, R.H. (1941). Dispersion and absorption in dielectrics, I: alternating current characteristics. *Journal of Chemical Physics*, **9**, 341–51.

Cornish, B.H., Ward, L.C. & Thomas, B.J. (1992a). Alteration of the extracellular and total body water volumes measured by multiple frequency bioelectrical impedance analysis (MFBIA). *Nutrition Research*, **14**, 717–27.

Cornish, B.H., Ward, L.C. & Thomas, B.J. (1992b). Measurement of extracellular and total body water of rats using multiple frequency bioelectrical impedance analysis. *Nutrition Research*, **12**, 656–66.

Cornish, B.H., Thomas, B.J. & Ward, L.C. (1993). Improved prediction of extracellular and total body water using impedance loci generated by multiple frequency bioelectrical impedance analysis. *Physical Medical Biology*, **38**, 337–46.

Cornish, B.H., Ward, L.C., Thomas, B.J., Jebb, S.A. & Elia, M. (1996). Evaluation of multiple frequency bioelectrical impedance and Cole–Cole analysis for the aseessment of body water volumes in healthy humans. *European Journal of Clinical Nutrition*, **50**, 159–64.

Cosgrove, J.R.J., King, W.B. & Brodie, D.A. (1988). A note on the use of impedance measurments for the prediction of carcass composition in lambs. *Animal Production*, **47**, 311–15.

Davies, P.S.W., Preece, M.A., Hicks, C.J. & Halliday, D. (1988). The prediction of total body water using bioelectrical impedance in children and adolescents. *Annals of Human Biology*, **15**, 237–40.

de Vries, P.M.J.M., Meijer, J.H., Oe, L.P., van Bronswijk, H., Schneider, H. & Donker, A.J.M. (1987). Conductivity measurements for analyses of transcellular fluid shifts during hemodialysis. *Transactions of the American Society of Artificial Internal Organs*, **33**, 554–6.

DeLorenzo, A., Andreoli, A., Matthie, J. & Whithers, P. (1997). Predicting body cell mass with bioimpedance by using theoretical methods: a technological review. *Journal of Applied Physiology*, **82**, 1542–58.

Deurenberg, P., van, de. K.K. & Leenen, R. (1989). Differences in body impedance when measured with different instruments. *European Journal of Clinical Nutrition*, **43**, 885–6.

Deurenberg, P., Weststrate, J. A., Paymans, I. & Kooy, K. v. d. (1988). Factors affecting bioelectrical impedance measurements in humans. *European Journal of clinical Nutrition*, **42**, 1017–22.

Drent, J., Marken Lichtenbelt, W. D. v. & Wikelski, M. (1999). Effects of foraging mode and season on the energetics of marine iguanas, *Amblyrhynchus cristatus*. *Functional Ecology*, **13**, 493–9.

Ducro, H., Thomasset, A., Joly, R., Jungers, F., Lenoir, J. & Eyraud, C. (1970). Détermination du volume des liquids extracellulaires chez l'homme par la mesure de l'impedance corporelle totale. *Presse Médicale*, **78**, 2269–72.

Earthman, C.P., Matthie, J.M., Reid, P.M., Harper, I.T., Ravussin, E. & Howell, W.H. (2000). A comparison of bioimpedance methods for detecting of body cell mass change in HIV infection. *Journal of Applied Physiology*, **88**, 944–56.

Ellis, K.J. & Wong, W.W. (1998). Human hydrometry: comparison of multifrequency bioelectrical impedance with H_2O and bromide dilution. *Journal of Applied Physiology*, **85**, 1056–62.

Eilsen, R., Siu, M-L., Pineda, O. & Solomons, N. W. (1993). Sources of variability in bioelectrical impedance determinations in adults. *Human Body Composition*. New York: Plenum Press.

Farley, S.D. & Robbins, C.T. (1994). Development of two methods to estimate body composition of bears. *Canadian Journal of Zoology*, **72**, 220–6.

Gallagher, M.R., Walker, K.Z. & O'Dea, K. (1998). The influence of a breakfast meal on the assessment of body composition using bioelectrical impedance. *European Journal of Clinical Nutrition*, **52**, 94–7.

Geddes, L.A. & Baker, L.E. (1967). The specific resistance of biological material: a compendium of data for the biomedical engineer and physiologist. *Medical and Biological Engineering and Computing*, **5**, 271–93.

Geerling, B.J., Marken Lichtenbelt, W.D.v., Stockbrügger, R.W. & Brummer, R-J.M. (1999). Gender specific alterations of body composition in patients with inflammatory bowel disease compared with controls. *European Journal of Clinical Nutrition*, **53**, 479–85.

Gray, D.S., Bray, G.A., Gemayel, N. & Kaplan, K. (1989). Effect of obesity on bioelectrical impedance. *American Journal of Clinical Nutrition*, **50**, 255–60.

Gudivaka, R., Schoeller, D. & Kushner, R. F. (1996). Effect of skin temperature on multifrequency bioelectrical impedance analysis. *Journal of Applied Physiology*, **81**, 838–45.

Hanai, T. (1968). Electrical properties of emulsions. *Emulsion Science*, pp. 354–77. London: UK: Academic.

Heitmann, B.L. (1990). Prediction of body water and fat in adult Danes from measurement of electrical impedance. A validation study. *International Journal of Obesity*, **14**, 789–802.

Heitmann, B.L. (1994). Impedance: a valid method in assessment of body composition? *European Journal of Clinical Nutrition*, **48**, 228–40.

Hildebrand, G.V., Farley, S.D. & Robbins, C.T. (1998). Predicting body condition of bears via two field methods. *Journal of Wildlife Management*, **62**, 406–9.

Hoffer, E. C., Meador, C. & Simpson, D.C. (1969). Correlation of whole-body impedance with total body water volume. *Journal of Applied Physiology*, **27**, 531–4.

Houtkooper, L.B., Going, S.B., Lohman, T.G., Roche, A.F. & Van, L.M. (1992). Bioelectrical impedance estimation of fat-free body mass in children and youth: a cross-validation study. *Journal of Applied Physiology*, **72**, 366–73.

Houtkooper, L.B., Lohman, T.G., Going, S.B. & Howell, W.H. (1996). Why bioelectrical impedance analysis should be used for estimating adiposity. *American Journal of Clinical Nutrition*, **64**, 436S-48S.

Jaffrin, M.Y., Maasrani, M., Le Gourrier, A. & Boudailliez, B. (1997). Extra- and intracellular volume monitoring by impedance during haemodialysis using Cole–Cole extrapolation. *Medical and Biological Engineering and Computing*, **35**, 266–70.

Janssen, Y.J., Deurenberg, P. & Roelfsema, F. (1997). Using dilution techniques and multifrequency bioelectrical impedance to assess both total body water and extracellular water at baseline and during recombinant human growth hormone (GH) treatment in GH-deficient adults. *Journal of Clinical Endocrinology and Metabolism*, **82**, 3349–55.

Jenin, P., Lenoir, J., Roullet, C., Thomasett, A.L. & Ducrot, H. (1975). Determination of body fluid compartments by electrical impedance measurements. *Aviation and Space Environmental Medicine*, **46**, 152–5.

Kanai, H., Haeno, M. & Sakamoto, K. (1987). Electrical measurement of fluid distribution in legs and arms. *Medical Progress Technology*, **12**, 159–70.

Khaled, M.A., McCutcheon, M.J., Reddy, S., Pearman, P.L., Hunter, G.R. & Weinsier, R.L. (1988). Electrical impedance in assessing human body composition: the BIA method. *American Journal of Clinical Nutrition*, **47**, 789–92.

Kreel, B.K. v., Cox-Reyven, N. & Soeters, P. (1998). Determination of total body water by multifrequency bio-electric impedance: development of several models. *Medical and Biological Engineering and Computing*, **36**, 337–45.

Kushner, R.F., Gudivaka, R. & Schoeller, D.A. (1996). Clinical characteristics influencing bioelectrical impedance analysis measurements. *American Journal of Clinical Nutrition*, **64**(S), 423S–7S.

Kushner, R.F. & Schoeller, D.A. (1986). Estimation of total body water by electrical impedance analysis. *American Journal of Clinical Nutrition*, **44**, 417–24.

Liang, M.Y. & Norris, S. (1993). Effects of skin blood flow and temperature on bioelectric impedance after exercise. *Medical Science Sports Exercise*, **25**, 1231–9.

Lukaski, H.C., Johnson, P.E., Bolonchuk, W.W. & Lykken, G.I. (1985). Assessment of fat free mass using bioelectrical impedance measurements of the human body. *American Journal of Clinical Nutrition*, **41**, 810–17.

Lundvall, J., Bjerkhoel, P., Quittenbaum, S. & Lindgren, P. (1996). Rapid plasma volume decline upon quiet standing reflects large filtration capacity in dependent limbs. *Acta Physiologica Scandinavica*, **158**, 161–7.

Marcello, M.J. & Slanger, W.D. (1992). Use of bioelectrical impedance to predict leanness of Boston butts. *Journal of Animal Science*, **70**, 3443–50.

Marcello, M.J. & Slanger, W.D. (1994). Bioelectrical impedance can predict skeletal muscle and fat-free skeletal muscle of beef cows and their carcasses. *Journal of Animal Science*, **72**, 3118–23.

Marken Lichtenbelt, W.D.v. & Fogelmholm, M. (1999a). Body composition. *Regulation of Food Intake and Energy Expenditure*, pp. 383–404. Milano: EDRA.

Marken Lichtenbelt, W.D.v. & Fogelmholm, M. (1999b). Increased extracellular water compartment, relative to intracellular water compartment, after weight reduction. *Journal of Applied Physiology*, **87**, 294–8.

Marken Lichtenbelt, W.D.v., Wesselingh, R.A., Vogel, J.T. & Albers, K.B.M. (1993). Energy budgets in free-living green iguanas in a seasonal environment. *Ecology*, **74**, 1157–72.

Marken Lichtenbelt, W.D.v., Westerterp, K.R., Wouters, L. & Luijendijk, S.C.M. (1994). Validation of bioelectrical-impedance measurements as a method to estimate body-water compartments. *American Journal of Clinical Nutrition*, **60**, 159–66.

Marken Lichtenbelt, W.D.v., Snel, Y.E.M., Brummer, R-J.M. & Koppeschaar, H.P.F. (1997). Deuterium and bromide dilution, and bio-impedance spectrometry independently show that growth hormone deficient adults have an enlarged extracellular water compartment related to intracellular water. *Journal of Clinical Endocrinology and Metabolism*, **82**, 907–11.

Matthie, J.R. & Withers, P.O. (1996). Bioimpedance, the Cole model equation and the prediction of intra- and extracellular water: science or marketing. *Clinical Nutrition*, **15**, 147–9.

Matthie, J.R., Withers, P.O., Van Loan, M.D. & Mayclin, P.L. (1992). Development of a commercial complex bio-impedance spectroscopic (CBIS) system for determining intracellular water (ICW) and extracellular water (ECW) volumes. *Proceedings of the VIIIth International Conference on Electrical Bio-Impedance*, Kuopio, Finland.

Maw, G.J., MacKenzie, I.L. & Taylor, N.A. (1995). Redistribution of body fluids during postural manipulations. *Acta Physiologica Scandinavica*, **155**, 157–63.

Mayfield, S.R., Uauy, R. & Waidelich, D. (1991). Body composition of low-birth-weight infants determined by using bioelectrical resistance and reactance [see comments]. *American Journal of Clinical Nutrition*, **54**, 296–303.

Nyboer, J. (1959). *Electrical Impedance Plethysmography: The Electrical Resistive Measure of the Blood Pulse Volume*. Springfield, IL: CC Thomas.

Nyboer, J. (1981). Percent body fat by four terminal bio-electrical impedance and body density in college freshmen. *Proceedings of the Vth International Conference on Electrical Bio-Impedance*, Business for Academic Societies, Japan.

Oldham, N.M. (1996). Overview of bioelectrical impedance analyzers. *American Journal of Clinical Nutrition*, **64**(S), 405S–12S.

Organ, L.W., Bradham, G.B., Gore, D.T. & Lozier, S.L. (1994). Segmental bioelectrical impedance analysis: theory and application of a new technique. *Journal of Applied Physiology*, **77**, 98–112.

Patterson, R., Ranganathan, C., Engel, R. & Berkseth, R. (1988). Measurment of body fluid volume change using multiple impedance measurements. *Medical and Biological Engineering and Computing*, **26**, 33–7.

Pencharz, P.B. & Azcue, M. (1996). Use of bioelectrical impedance analysis measurements in the clinical management of malnutrition. *American Journal of Clinical Nutrition*, **64**(S), 485S–8S.

Pietraszek, J. & Atkinson, S. (1994). Concentrations of estrogen sulfate and progesterone in plasma and saliva, vaginal cytology, and bioelectric impedance during the estrous cycle of the Hawaiian Monk Seal (*Monochus schauinslandi*). *Marine Mammal Science*, **10**, 430–41.

Raghavan, C.V., Super, D.M., Chatburn, R.L., Savin, S.M., Fanaroff, A.A. & Kalhan, S.C. (1998). Estimation of total body water in very-low-birth-weight infants by using anthropometry with and without bioelectrical impedance and $H_2[(18)O]$. *American Journal of Clinical Nutrition*, **68**, 668–74.

Rising, R., Swinburn, B., Larson, K. & Ravussin, E. (1991). Body composition in Pima Indians: validation of bioelectrical resistance. *American Journal of Clinical Nutrition*, **53**, 594–8.

Rodríguez, G., Moreno, L. A., Sarría, A., Fleta, J. & Bueno, M. (1999). Diurnal variation in the assessment of body composition using bioelectrical impedance in children – letter to the editor. *European Journal of Clinical Nutrition*, **53**, 244.

Rowell, L.B. (1993). Reflex control curing orthostasis. *Human Cardiovascular Control*, pp. 37–80. New York: Oxford University Press.

Scharfetter, H., Wirnsberger, G., Laszzlo, Z., Holzer, H., Hinghofer-Salkay, H. & Hutten, H. (1995). Influence of ionic shifts and postural changes during dialysis on volume estimation with multifrequency impedance analysis. *Proceedings of the IXth International Conference on Electrical Bio-impedance*. Heidelberg, University of Heidelberg.

Scheltinga, M.R., Jacobs, J.O., Kimbrough, T.D. & Wilmore, D.W. (1991). Alterations in body fluid content can be detected by bioelectrical impedance analysis. *Journal of Surgical Research*, **50**, 461–8.

Schoeller, D.A. (1996). Hydrometry. In *Human Body Composition*, pp. 25–44. Leeds: Human Kenetics.

Schwan, H.P. (1957). Electrical properties of tissue and cell suspension. *Advances in Biological and Medical Physics*, pp. 149–209. New York: Academic.

Segal, K.R., Van Loan, M.D. & Fitzgerald, P.I. (1988). Lean body mass estimated by bioelectrical impedance analysis: a four site cross validation study. *American Journal of Clinical Nutrition*, **47**, 7–14.

Settle, R.G., Foster, K.R., Epstein, B.R. & Mullen, J.L. (1980). Nutritional assessment: whole body impedance and body fluid compartments. *Nutrition and Cancer*, **2**, 72–80.

Smith, D.N. (1993). Biolelectrical impedance and body composition. *Lancet*, **341**, 569–70.

Smye, S.W., Norwood, H.M., Buur, T., Bradbury, M. & Brocklebank, J.T. (1994). Comparison of extra-cellular fluid volume measurement in children by 99Tcm-DPTA clearance and multi-frequency impedance techniques [see comments]. *Physiological Measurement*, **15**, 251–60.

Steen, B., Bosaeus, S., Galvard, H., Isaksson, B. & Robertson, E. (1987). Body composition in the elderly estimated with an electrical impedance method. *Comparative Gerontology*, **1**, 102–5.

Stroud, D.B., Cornish, B.H., Thomas, B.J. & Ward, L.C. (1995). The use of Cole–Cole plots to compare two multi-frequency bioimpedance instruments. *Clinical Nutrition*, **14**, 307–11.

Thomasset, A. (1962). Bio-electrical properties of tissue impedance measurments. *Lyon Medical*, **207**, 107–18.

Thomson, B.C., Thomas, B.J., Ward, L.C. & Sillence, M.N. (1997). Evaluation of multifrequency bioelectrical impedance data for predicting lean tissue mass in beef cattle. *Australian Journal of Experimental Agriculture*, **37**, 743–9.

Van Loan, M. D., Withers, P., Matthie, J. & Mayclin, P. L. (1993). Use of bioimpedance spectroscopy to determine extracellular fluid, intracellular fluid, total body water, and fat-free mass. *Human Body Composition*, pp. 67–70. New York: Plenum Press.

Van Loan, M.D., Kopp, L.E., King, J.C., Wong, W.W. & Mayclin, P.L. (1995). Fluid changes during pregnancy: use of bioimpedance spectroscopy. *Journal of Applied Physiology*, **78**, 1037–42.

Wang, Z., Heshka, S., Pierson, R.N. Jr & Heymsfield, S.B. (1995). Systematic organisation of body-composition methodology: an overview with emphasis on component-based methods. *American Journal of Clinical Nutrition*, **61**, 457–65.

Wang, Z., Deurenberg, P., Wang, W., Pietrobelli, A., Baumgartner, R.N. & Heymsfield, S.B. (1999). Hydration of fat-free body mass: review and critique of a classic body-composition constant. *American Journal of Clinical Nutrition*, **69**, 833–41.

Westerterp, K.R., Marken Lichtenbelt, W.D. v. & Wouters, L. (1995). The Maastricht protocol for the measurement of body composition and energy expenditure with labeled water. *Obesity Research*, **3**, 49–57.

Woolnough, A.P., Foley, W.J., Johnson, C.N. & Evans, M. (1997). Evaluation of techniques for indirect measurement of body composition in a free-ranging large herbivore, the Southern Hairy-nosed Wombat. *Wildlife Research*, **24**, 649–60.

Zhu, F., Schneditz, D., Wang, E. & Levin, N.W. (1998). Dynamics of segmental extracellular volumes during changes in body position by bioimpedance analysis. *Journal of Applied Physiology*, **85**, 497–504.

J. MATTHIAS STARCK, MAURINE W. DIETZ
AND THEUNIS PIERSMA

7

The assessment of body composition and other parameters by ultrasound scanning

Introduction

Ultrasound has been employed in various technologies. It is a tool for civil and military marine location and navigation (SONAR, Sound Navigation and Ranging), and is also used for quality control in material sciences, and as a non-invasive tool for diagnostic imaging in human and veterinary medicine. Evolution has made use of ultrasonic physical phenomena, and many animals use ultrasound for orientation and communication, e.g. bats, whales, small mammals, some birds.

The physical principle underlying all ultrasound applications, technical and biological, is basically the same: a very high frequency sound is emitted from a source and travels some distance through a medium until it hits a reflector that returns it as an echo to its source. Time differences between sound emission and the returning echo, and the intensity of the echo may be used to obtain information about distance and about the properties of the reflector. It depends on the computational procedure whether the echo information is employed to give local information, navigational cues, or to produce images.

This chapter is about imaging, and we leave ranging and navigation for other discussions. Ultrasound imaging was developed in the 1950s to provide a diagnostic tool in human medicine (Morneburg, 1995). Much progress has been made since the late 1960s, when advanced algorithms and computer technologies allowed for faster data processing and real time imaging. Today, ultrasound imaging is a pivotal diagnostic tool in almost all fields of human medicine, and has gained increasing importance in veterinary medicine and reproductive management (Göritz, 1993; Hildebrandt *et al.*, 1998). However, it is still a rather rare tool in zoological and ecological research.

Theory

The basics

Human auditory sensitivity has an upper limit at about 20 kHz. Higher frequencies are classified as ultrasound. Physically, sound is a mechanical disturbance of the thermodynamic balance of materials which is propagated in waves. The velocity of acoustic waves depends on the density of the material, e.g. sound velocity ranges between 331 m s^{-1} in air and 3500 m s^{-1} in bone. Acoustic waves can be reflected, refracted, diffracted, and absorbed. In liquids and gases sound is propagated primarily in a longitudinal manner. Because the propagation of sound depends on material, no sound can be transmitted in a vacuum. Density differences between materials may act as reflectors, i.e. produce echoes. Marked density differences between two adjacent materials cause full reflection, e.g. the water to air barrier reflects 99.9% of the sound energy. As a consequence, such density differences cannot be bridged by ultrasound and cause significant limitations for ultrasound imaging.

The 'heart' of the ultrasound imaging technique is a piezoelectric crystal. Appropriate electric stimulation of the crystal causes emission of a short series of ultrasound pulses which propagate longitudinally into the tissue. Density differences in the tissue reflect part of the ultrasound pulse, which returns as an echo to the probe. During the intervals between transmissions from the crystal, returning echoes cause a mechanical deformation of the crystal and thus induce an electric current that can be amplified and measured as the echo signal. The velocity of the ultrasound in the particular tissue, and the time lapse between pulse emission and the returning echo, render the information necessary to calculate the distance between the probe and the reflector (A-mode; amplitude modulation). If, in addition, an area is scanned by moving the crystal over the tissue and if the intensity of returning echoes is coded in grey levels, a two-dimensional image can be produced in which the grey levels encode the intensity of the echoes (the echogeneity of the tissues; B-mode or brightness modulation). In B-mode imaging, tissues with the same echogeneity will appear in the same grey level (Fig. 7.1). However, diffraction and absorbance of ultrasound cause a significant loss of sound energy. Therefore, echoes returning from deep reflectors will arrive not only later but also weaker than those from closer reflectors. Usually, such depth effects are compensated electronically by run-time dependent amplification of echoes, e.g. log scale amplification; time gain control TGC.

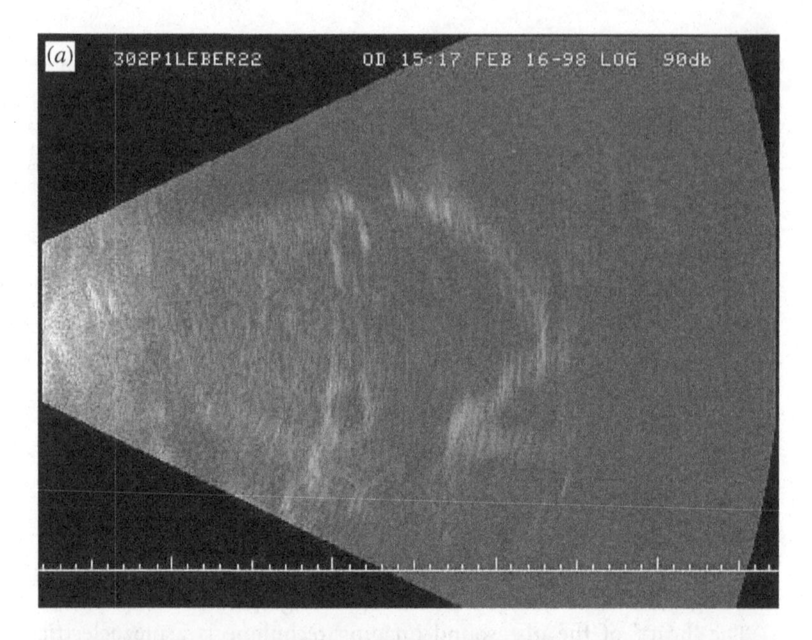

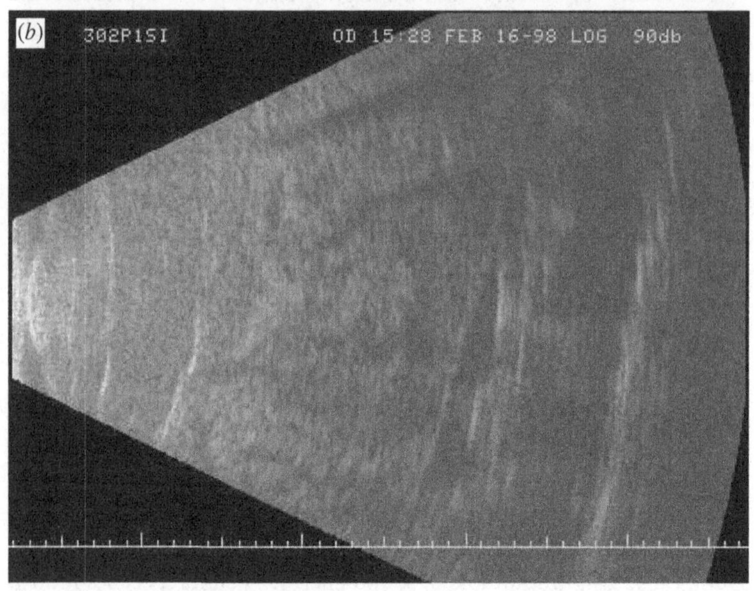

Fig. 7.1. B-Mode imaging in ultrasonography. (*a*) Ultrasonograph of a 'cross-section' the liver of a Burmese python (10 MHz, sector scanner). (*b*) Ultrasonograph of the small intestine of a Burmese python.

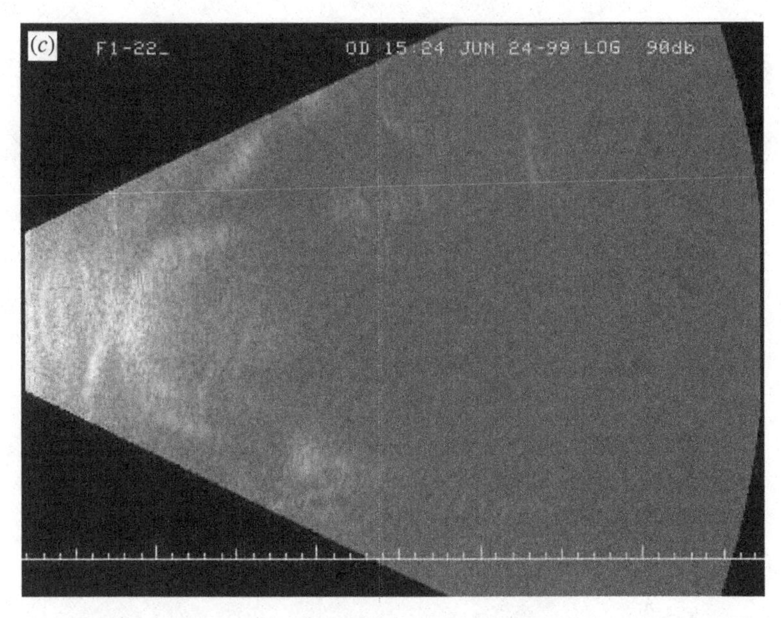

Fig. 7.1 (*cont.*) (*c*) Ultrasonograph of the muscular stomach of a Japanese quail. All images were taken using a 10 MHz, sector scanner.

For any application of ultrasonography, the resolution, i.e. the minimum distance at which two points can be distinguished as separate, is a central issue. From a physical point of view, a structure must be larger than the wavelength to reflect ultrasound waves. Otherwise, the wave will go around the object without reflection and no echo will occur.

Wavelength and sound frequency are related by:

$$\lambda = \frac{c}{\nu} \tag{1}$$

(λ = wavelength, ν = frequency, c = velocity). Consequently, the possible resolution (Δx) is determined by the wavelength, i.e. frequency, of the ultrasound.

$$\Delta x \geq \lambda \tag{2}$$

In tissues, c is on average around 1568 m s^{-1} (Hassler, 1995). Thus, a frequency of 1 MHz (10^6 Hz) has a wavelength of 1.5 mm, with according resolution. High frequencies give better resolution than low frequencies. In technical solutions for ultrasound imaging, the spatial resolution has two

Table 7.1. *Typical resolutions and image depth in ultrasound imaging in tissues*

Frequency (MHz)	Resolution (mm)		Image depth (mm)
	Lateral	Axial	
3.5	1.7	0.50	140
7.5	0.8	0.25	67
10	0.6	0.20	50
15	0.4	0.15	33
20	0.2	0.10	14

components, one along the ultrasound beam (axial resolution) and one perpendicular to the beam (lateral resolution = Δx). The axial resolution (Δz) depends on the run time difference between two succeeding echoes and is defined as:

$$\Delta z \geq \frac{\lambda}{2} \tag{3}$$

(Hassler, 1995). However, absorption, diffraction and reflection also depend on frequency. High frequency waves are more easily absorbed, diffracted and reflected than low frequencies. Consequently, high frequencies do not penetrate into tissues as deeply as low frequency waves. For the practitioner, this causes a trade-off between resolution and image depth. High frequency probes (>10 MHz) have a high resolution but only a limited image depth. Conversely, low frequency probes (<10 MHz) have lower resolution but the image depth is greater. Lateral and axial resolution, the technical properties of the scanner head, the peripheral computing facilities, and the image processing algorithms determine the effective resolution. Typical frequencies of ultrasound scanner heads, their resolutions, and their image depths are given in Table 7.1. In practice, an ultrasound machine will always be equipped with several scanner heads allowing one to use different frequencies to satisfy the specific needs of different applications.

Extended basics: the Doppler principle

If a reflector is moving within an acoustic field, the frequency of the reflected echoes changes according to the speed and direction of the object; this is the Doppler principle. The echo frequency increases when the reflector moves towards the ultrasound source, and decreases if the reflector moves away from the source. When ultrasound is emitted continuously (continuous wave doppler, CW-Doppler), frequency shifts can be used to measure the velocity of a moving reflector that is large enough to be detected within the reach of ultrasound waves. If the frequency of the returning echoes is mixed with the frequency of the emitted sound, the frequency difference is within the range of human auditory sensitivity and can be heard acoustically with a speaker, or be visualized using an oscilloscope. However, because the ultrasound wave is emitted continuously, no topographic localization of the moving object is possible with a CW-Doppler.

To obtain topographic information about a moving object, a pulsed ultrasound wave (PW-Doppler) is sent into the tissue. The time window between the pulses (pulse repetition frequency, PRF) provides information about the depth of the reflector. It can be adjusted according to the need of the investigation. The maximum speed of a moving object that can be measured is determined by the PRF. Modern ultrasound machines use colour coding (colour flow mode, CFM) to visualize velocity information of the PW-Doppler signal. When the PW-Doppler signal is overlain upon a B-mode image, information about the velocity of moving objects, e.g. blood in vessels can be combined with detailed topographic information. This is called Duplex-sonography. Recent developments allow for the additional measurement of the changes of amplitude of the PW-signal, and thus for the detection of very slow flow (Power-Doppler).

Technical aspects

Within the framework of physics, technical solutions determine the quality of ultrasound images. It is beyond the scope of this contribution to go into the details of the swiftly changing technical aspects of attributes such as ultrasound scanner heads and associated computer software. Nevertheless, we think it is important to briefly outline the basic features of current technology to provide some rough guidelines for those choosing an ultrasound system according to the specific needs of a research project.

Mechanical scanner heads have just one piezoelectrical crystal that is moved over the study object at a certain scanning angle. Compared to linear arrays (see below), the construction of a mechanical scanner head is 'simple', and the computational effort required to interpret the signals is low. The technology does not require much space, so small scanner heads can be made with scanning angles up to 360°. The image quality of mechanical scanner heads is generally high. The cost-effectiveness relationship is good, since an excellent resolution usually comes at a moderate price. However, because of the high inertia of the mechanical parts, switching between different modes, e.g. B-mode to CW-Doppler is not possible, and mechanical scanner heads cannot be used in Duplex-Sonography. Today, mechanical scanner heads have specialized applications in echocardiography (CW-Doppler), endo-sonography, endovascular sonography, and ophthalmology (A-mode, B-mode).

Electronic scanner heads are characterized by arrays of piezoelectric crystals arranged in a linear (linear arrays) or curved (convex arrays) manner. The standard is 196 elements in linear arrays. During each moment of operation, only a small subset of all elements sends pulses and receives echoes. Scanning proceeds by including elements on one side of the subset and turning off elements on the other side. The arrangement of neighbouring elements requires thorough mechanical isolation of parts, and the elaborate electronics demand extensive computing. Because no mechanical inertia is associated with the elements, switching between modes is easy and fast. Therefore, all modern Duplex sonographs are equipped with electronic scanner heads. Electronic scanner heads can be built in various sizes and forms. Most are linear or convex, but many other forms are possible for more specialized applications, e.g. intraoperative and fingertip probes, miniature probes for endoscopic ultrasonography.

Phased arrays represent a combination of electronic and mechanical scanners. Here, a small array, in which all elements are active at once, is mechanically scanned over the object. Typical equipment has 48–128 elements and a scanning angle of 40–45°. A few companies have started building true multifrequency scanner heads. Those emit a broad frequency band, e.g. 5–12 MHz and analyse returning echoes accordingly. Such scanner heads combine the high resolution of high frequencies with high image depth of lower frequencies. However, the computational effort required to analyse the echoes is enormous and usually requires powerful and extremely fast image processors.

Besides the construction of a scanner head, the dynamic range of an ultrasound system is the most important feature for high quality imaging. The dynamic range determines the width of the range of energy of the echoes that is used to build up an image. A wide dynamic range uses more information to form the image, and it retrieves echoes from deeper parts of the scanned area than a narrow dynamic range; the actual standard is

110 dB. Good systems have a dynamic range of 130 dB; some outstanding products have one of 150 dB. However, a wide dynamic range also requires extensive computing with consequently increasing costs.

Three-dimensional ultrasound imaging

Conventional ultrasound imaging produces two-dimensional images, comparable to anatomical sections through organs. For any understanding of more complex structures, detailed morphometric measurements, and for volumetric measurements, three-dimensional imaging is needed. Today, two different technical approaches introduce the third dimension in ultrasound imaging. (i) 3D-imaging produces real time 3D ultrasound images. In contrast to conventional two-dimensional ultrasonography, 3D imaging requires scanning of a volume and storage of data in a complex three-dimensional data matrix. It necessitates complex scanning heads and employment of powerful parallel processing computers to manage the amount of data. For example, the amount of data processed per second of real time 3D-imaging would fill 6000 CD ROMs. Presently, only two real-time 3D scanner heads are available with a limited range of frequencies and applications. (ii) A technically simpler solution has been invented by 3D reconstruction of spatially aligned two-dimensional images. 3D reconstruction works with conventional scanner heads and some additional equipment to provide information about the localization of the scanner head and the images. Two-dimensional images are stored together with spatial information, and are later reconstructed as a pile of images. Surface views, transects, and various viewing angles can later be computed using the reconstructed image pile. Three-dimensional reconstruction of images requires 2D image acquisition, storage of the images, together with their topographic information, and finally alignment of images to form the reconstruction. All steps are interactive and need programming by the user. Therefore, extra time is always necessary for calculating the reconstruction and one will never see immediate results from a scan.

However, 3D images may also be obtained with standard equipment and relatively low computing effort. Usually, conventional personal computer technology suffices to achieve high quality reconstructions. Several systems are available that require the necessary localization information of the scanner head for the correct alignment of images during reconstruction. Such information is gathered either by moving the scanner head within a solid frame positioned over the object, or by moving the scanner in an external magnetic field and recording its position in the

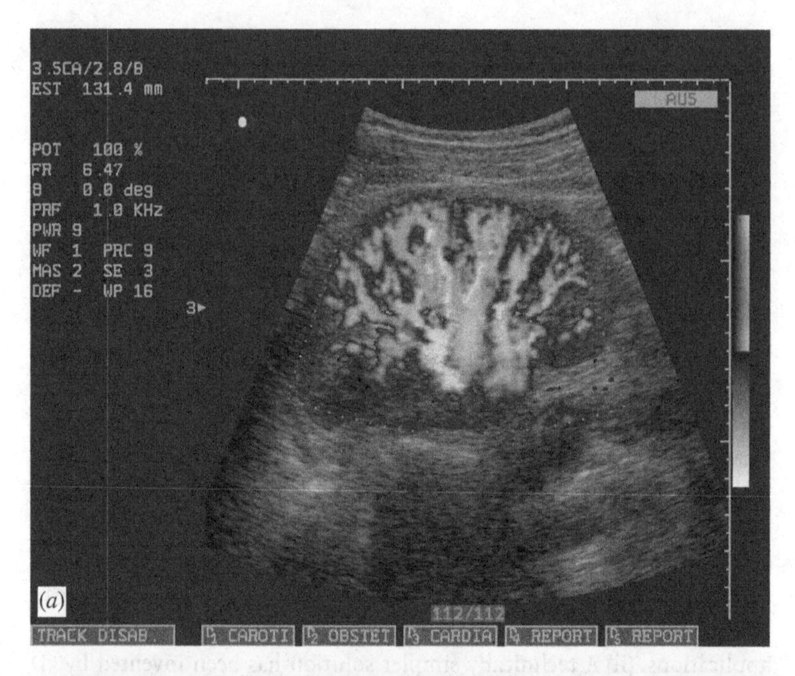

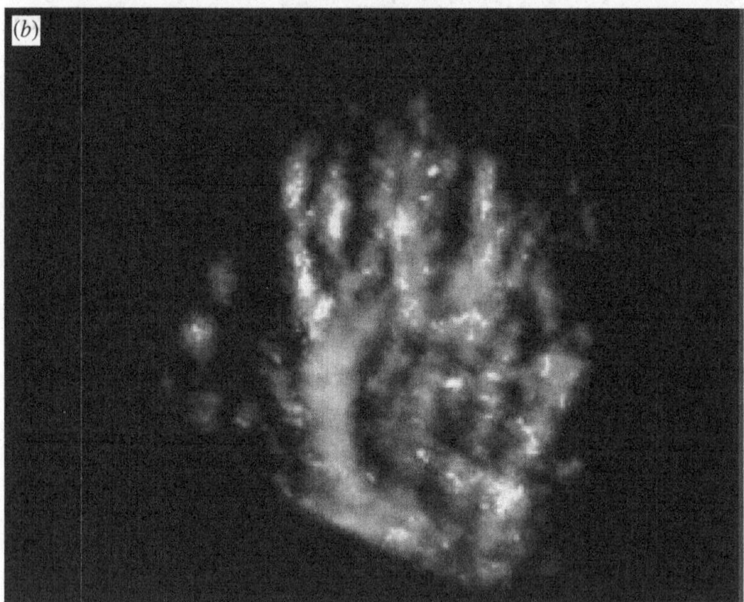

Fig. 7.2. Duplex sonography and 3D imaging. (*a*) Bidirectional colour coding of blood flow in the human kidney (courtesy of Esaote Ltd). (*b*) 3 D reconstruction of blood flow pattern in the human kidney (courtesy of Esaote Ltd).

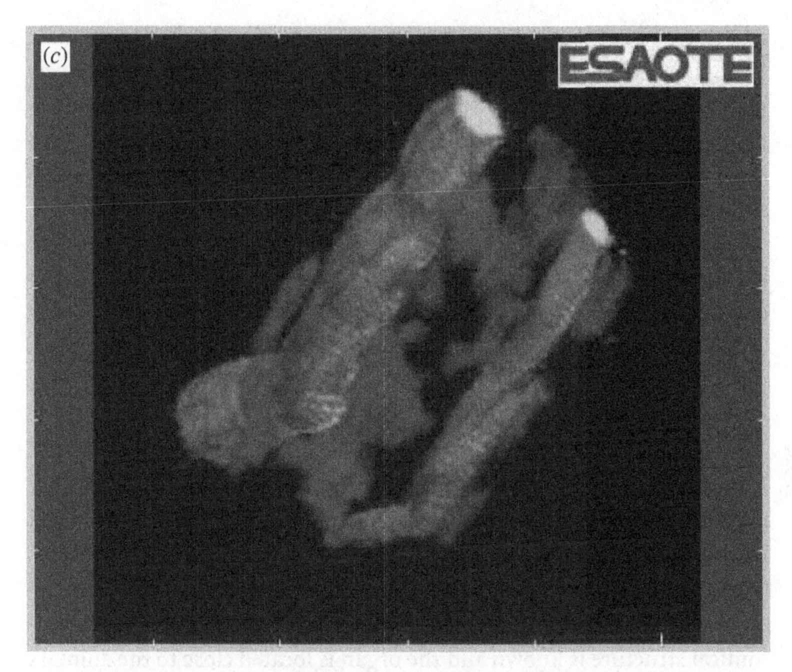

Fig. 7.2 (*cont.*) (*c*) 3D reconstruction of the liver portal vein and the liver vein in the liver of a Burmese python. (Unpublished material J.M. Starck.)

field. Limitations come through original image quality, image processing, potentially restricted image storing capacities, and time necessary to reconstruct images. Recent solutions are promising (see Fig. 7.2), but the field of 3D reconstructions changes rapidly and future developments will certainly further improve 3D reconstruction.

Practical aspects

General

Ultrasound scanning has the following advantages compared to imaging techniques like radiography and nuclear magnetic resonance: (i) ultrasound is completely non-invasive, (ii) it allows for real-time imaging, (iii) it allows the detection and measurement of speed and direction of motion, e.g. heart rate, vascular flow, and fetal movements, (iv) it allows morphometric measurements of structures *in situ*, e.g. organs and implants, and (v) it facilitates documentation of digital data on storable media, e.g. floppy disks and video so that images can be processed later. In addition, (vi) some ultrasound systems are portable and thus compatible with both laboratory and field applications, and (vii) they are relatively inexpensive and simple to use and do not need much specialized technical training (Hildebrandt *et al.*, 1998). Ultrasound, by itself, causes no discomfort (see below), and because manual restraint of the animal is often sufficient, it is also a relatively non-stressful method.

Ultrasound scanning can be used for a variety of purposes, each with their own possibilities and limitations. Table 7.2 summarizes the most commonly used applications for the different taxa. Although we could find only one study for both molluscs and crustaceans, other applications are likely to be possible in these taxa as well.

Procedures

Although ultrasound scanning is relatively easy and simple, just as other non-invasive imaging techniques, it does require detailed anatomical knowledge and experience when it comes to the interpretation of images (Starck & Burann, 1998). The echogenetic properties of the organ of interest and the surrounding tissues should be known. Even when the anatomical structure is known and the organ is located close to the animal's body surface, topographic localization of that organ may still present a problem when, for example, variable amounts of fat may affect the image quality. The location of an organ might differ between individuals as well as within individuals. In shorebirds such as red knots (*Calidris canutus*), the stomach is conveniently located at the centre of the abdomen when it is large, but stomachs usually lie at the left side of the abdomen when small. Finding such small stomachs is rather difficult with a flat probe, because the abdomen is round. Also, small stomachs have a tendency to disappear underneath airsacs. Observers therefore, not only need anatomical knowledge and experience with the interpretation of the images, they need patience as well.

Ultrasound imaging provides 'acoustic sections' through (parts of) an animal. Considering the scanner head as an acoustic knife: the positioning of the probe on the surface of the animal determines the plane and direction of the image. Even minute tilting of the scanner head may create considerably different image contents. Therefore, clearly defined and reproducible positions of the scanner head on the surface of an animal are a necessary prerequisite to produce reproducible, and thus, comparable images. Surface topography has been used for many medical and veterinary applications and a specific nomenclature that describes the positioning of the probe has been developed. For many exotic animals, unusual positions of the probe for transcutaneous ultrasound imaging may be necessary. A precise description of the positioning of the probe onto the animal is imperative for any later replication of the study.

In transcutaneous ultrasonography, the use of a water-based contact gel between the probe and the animal is required, to ensure that air layers do not destroy the image, except for transcutaneous ultrasonography of animals that are submerged in water. Bird species that have clear unfeathered areas (apteria) at appropriate places are easy to work with. For example, many birds species have a bare patch of skin on the abdomen, just caudal of the keel. Often, this bare spot can be moved to the location of interest. In species in which the density of the feathers is low, the feathers can be wetted and rubbed to the side before the application of contact gel. The coupling gel can be washed off with lukewarm water, and in our experience with shorebirds even scanning at 2-day intervals does not cause plumage problems. In waterbirds such as divers, grebes, ducks and swans the dense layer of feathers provides a real obstacle to the use of ultrasonography. This is also true of animals with a pelage, where the dense layer of hair needs to be clipped or shaved off before contact gels can be applied and scanning can proceed. When the removal of hair is not an option, soaking the fur with alcohol can help (Hildebrandt *et al.*, 1998).

Transcutaneous scanning is probably the most commonly used application of ultrasound. Alternatively, transintestinal, transrectal, transcloacal or transoesophageal ultrasonography may be performed when anatomical properties of the animal prohibit transcutaneous ultrasonography (Hildebrandt *et al.*, 1998). These applications allow a better imaging of the abdominal structures and may be the only option when airsacs, thick scales, a very thick fur, or plumage prevents transcutaneous scanning. However, such semiinvasive forms of ultrasonography usually require that the animal is securely restrained or sedated, which in itself causes some stress. Also, when entering animals through the anus, faeces have to be removed before scanning for example by a lavage with warm water (Hildebrandt *et al.*, 1994, 1995). Endosonography requires specially designed and miniaturized probes.

Imaging artefacts

The use of ultrasonography is limited by the skeleton and airsacs (in birds, fish and some snakes), because both air and bones reflect the ultrasound signal which prevents the imaging of structures underneath (Starck &

Table 7.2. *Main applications of ultrasound in animal classes*

Class	Applications	Probe	Coupling gel	Scanning problems	Restraint or sedation
Molluscs	Heart rate (muscles were also visible)	7.5 MHz linear array	Not needed when animal remains in water	Not for this purpose	Forced restraint, securing animal to surface
Crustaceans	Heart rate Scaphognathite rate	7.5 MHz linear array	Not needed when animal remains in water	Not for these purposes	Forced restraint, securing the animal to the surface
Fish	Sex determination Reproductive organ development Muscle development Diagnostic applications	Small fish: 7.5–10.0 MHz Large fish: 3.5–5.0 MHz	Not needed when animal remains in water	Hard and thick scales will interfere with ultrasound waves; transintestinal scanning may overcome these problems	Small fish are usually sedated, for larger fish manual restraints sufficient
Amphibians	Sex determination Diagnostic applications Reproductive organ development Anatomical studies	5.0–10.0 MHz	Not needed when animal remains in water	None	Sedation, manual restraints or cooling by placing animal in an ice water filled container
Reptiles	Sex determination Reproductive organ development Egg development Diagnostic applications Anatomical studies Digestive studies	Small animals: 7.5–10.0 MHz Large animals: 3.5–5.0 MHz	Apply freely	Animals are best scanned from ventral or lateral surface. Hard and thick scales will interfere with ultrasound waves; trans-intestinal scanning may overcome these problems	Usually manual restraint is sufficient, preferably while animals are resting on their sternum e.g. on platforms with holes for the probe
Birds	Muscle evaluation Sex determination Reproductive organ development	5.0–10.0 MHz	Apply freely, can easily be removed from the feathers with water	Transcutaneous applications are relatively limited due to feathers and down, fat around feather follicles,	Usually manual restraint is sufficient

	Eggshell measurements Diagnostic applications Ontogenetic studies Digestive studies Anatomical studies			airsacs and sometimes by long caudally placed keels Most transcutaneous scanning takes place via the ventral midline, just below the keel. Transintestinal scanning may overcome some problems	
Mammals	Muscle evaluation Fat mass determination Sex determination Reproductive organ development Pregnancy detection Diagnostic applications Ontogenetic studies Digestive studies Anatomical studies	3.5–10.0 MHz depending on size of the animal	Apply freely, but not needed with marine animals submerged in water.	Hairs should be clipped from site of interest. When that is not acceptable, the area may be soaked with alcohol Transendosonography is possible, especially transrectal scanning	Depending on animal and location of scanning manual restraint could be sufficient, otherwise forced restraint or sedation required

Source: Haefner, 1996; Haefner *et al.*, 1996; Hildebrandt *et al.*, 1998.

Burann, 1998). Also, the filling of the stomach and or the intestines may interfere with the image. Impermeable small structures such as ribs may give 'shadow' artefacts; the shadow appears separately underneath the actual structure (Starck & Burann, 1998). Movement artefacts, caused by movement of the whole animal or some of its tissues, can be troublesome. This is overcome by storing images on video. Another solution for this problem is to store sequences of images (currently different scanners are able to retain 10–170 s of image), which can then be played back on the scanner itself. In this way the best image can be selected.

The physical properties of ultrasound imaging inevitably lead to some image artefacts. Understanding these artefacts is of utmost importance when applying ultrasonography. For example, repetitive echoes (reverberations) may occur when the ultrasound echoes are repeatedly reflected between tissues of relatively high density difference. For example, metallic or glass objects and air (between probe and skin or within the organism) can cause reverberation artefacts. The typical appearance is a comet-tail like shadow attached to the cause and crossing tissue boundaries (Marcelis *et al.*, 1996). When ultrasound travels through a structure with very low absorbance, e.g. water, fluid filled cyst compared with surrounding tissue, the tissue behind that structure may appear with an unexpected bright echo; this is called dorsal amplification. This particular artefact can be used as a diagnostic tool for the determination of cyst contents. Shadow artefacts can occur when ultrasound waves hit an object that reflects all ultrasound energy, e.g. bone or air, and ultrasound waves cannot further penetrate the object. Shadow artefacts may also occur when all ultrasound energy is absorbed within a structure, e.g. bone.

Recording and documentation of ultrasonographs

Various procedures are available to record ultrasonographs and store images. Most clinical systems provide a video printer for printing the actual scan. Also, many systems offer digital storage of images within the system. However, practitioners in biological science will probably request more detailed and full records of ultrasound sessions and prefer to rely on continuous video tape recording of sessions. Also, one may wish to export stored images to other image processing facilities for later morphometry or exchange with other laboratories. Therefore, network capability and free exchangeable image formats, e.g. TIF-, or GIF-format images may become important aspects when ultrasonography is routinely used in a research laboratory.

Ultrasound impact on tissues and safety standards

Ultrasound may potentially affect tissues by locally raising the temperature and/or by cavitation effects, i.e. sound pressure differences causing microscopic disruptions of tissue. Both effects depend on the intensity of the ultrasound application. Thermal effects also depend on the duration of ultrasound application. No clear threshold has been recognized at which damage begins to occur in tissue. However, international medical safety standards have agreed on 100 mW cm^{-2} and unlimited treatment (continuous wave) as the upper limit (Soldner, 1995). In routine medical ultrasonography, the energy applied to tissues ranges between 1–10 mW cm^{-2}, and is thus about two orders of magnitude below the threshold for unlimited treatment. Also, one has to keep in mind that to keep axial resolution high the duration of the ultrasound pulse is very short, about 1 ms. Thus, even if a patient should experience the unusual treatment of 1 hour of permanent ultrasonography at an unchanged position, the exposure of tissues to ultrasound would be only 3.6 s. Physical experiments showed that cavitation effects occur when local sound pressure increases above 15 MPa. Under clinical conditions, medical scanner heads with frequencies of 2–20 MHz produce maximal pressures betweeen 0.5 and 1.5 MPa, thus giving a tenfold safety margin.

Pulsed wave and continuous wave Doppler ultrasonography applies ultrasound pulses in high frequency or continuously. Therefore, the energy input into the tissue is higher than with conventional B-mode imaging, and temperature effects cannot be excluded. No information is available to us that shows effects of a local increase in tissue temperature. However, with more than 20 years of prenatal diagnostics, other medical diagnostic treatments, and an increasing knowledge of veterinary applications, one may safely assume that endothermic organisms may efficiently deal with local increases of temperature under routine treatment. Nothing is known about how ectothermic animals respond to a potential local increase in tissue temperature. Therefore, it is advisable to apply the lowest possible energy, i.e. shortest possible treatment. Safety rules presently restrict application of CW- or PW-sonography in early human pregnancy, but no further clinical restrictions are known.

Validation studies

Repeatability of measurements

Exact knowledge of the internal morphology and the echoscopic features of the organ, i.e. the appearance of the organ on the ultrasound image is mandatory for a correct interpretation of the ultrasound image. If topographic anatomical information cannot be retrieved from the literature dissections of specimens are required. Satisfactory morphological knowledge can be obtained from frozen specimens that are sectioned into slices. In animals with airsacs, these should be inflated before the bodies are frozen. The echogenic features of an organ of interest can be tested by dissecting it out and submersing it in water. Then, the tissue can be scanned without coupling gel (Hildebrandt *et al.*, 1998; Starck & Burann, 1998). Ultrasound imaging, especially for morphometric measurement, need careful (i) verification of the results with conventional dissection, and (ii) assessment of the repeatability of the measurements (Hildebrandt *et al.*, 1998; Starck & Burann, 1998; Dietz *et al.*, 1999a). Up to now, the ultrasonographic appearance of different organs has only been published for a few taxa.

The plane of the ultrasound scanner (the 'acoustic section') and the repeatability of ultrasound measurements depend on surface positioning, angle of the head, and the pressure applied to the subject's skin. To determine the precise topographic surface position and angle of the scanner, anatomical 'landmarks' on the surface of the animal can be most helpful. For example, in birds the *incisura lateralis sterni* can be used to position the scanner head in relation to the sternal axis. Often, however, animals do not show helpful anatomical gridlines and this may require the application of (semi-) permanent skin marks to enhance the repeatability of the measurement.

The amount of pressure applied by the probe on the tissue influences the image of that organ. Tissue thickness and organ size will decrease with increasing pressure, but also the fine detail of the tissue, such as the amount of layers that are visible in the gastrointestinal wall, vary with the pressure applied (Ødegaard *et al.*, 1992). In principle, no pressure should be applied with the probe. Ødegaard *et al.* (1992) looked in detail at the effects of applied pressure on *in vitro* specimens of the porcine gastrointestinal tract (Advanced Technology Laboratories, 8.5 MHz linear array probe, real-time). Wall thickness decreased rapidly with pressure. At a pressure of only 1 kPa (~10.4 g/cm^2) the thickness of the stomach decreased by about 25%, while the thickness of softer tissues such as the walls of the oesophagus, duodenum, colon and rectum decreased by about 35–40%. According to Ødegaard *et al.* (1992) the maximal pressure that might be applied *in vivo* ranges between 6 and 19 kPa. However, in contrast to the *in vitro* situation, organs *in vivo* generally rest on other soft, movable tissues and the pressure applied may have less effect than *in vitro*. In addition, in most ultrasonographic applications, the observer can actually see what happens, and the pressure can therefore be relaxed if the image conveys distortion.

Calibration

Location, angle and pressure applied by the probe are all in the hand of the observer who also has to interpret the image and thus strongly influences the result, e.g. Urbak *et al.*, 1998. As the observer effect may decrease with increasing experience (Perkins *et al.*, 1992; Herring *et al.*, 1994) good repeatability can be obtained if observers are carefully trained. The training may be especially important in studies of structures that are not so clearly defined on the image (Urbak *et al.*, 1998). Calibration curves are specific to the observer, the species, the organ and the aim of the visualization. Although the calibration curve should ideally be made on live animals that are killed after the measurements, dead animals can also be used (Starck & Burann, 1998; Dietz *et al.*, 1999a). Note, however, that decay changes the structure and echogenicity of the tissues, so that any such calibrations may differ from those with live animals.

A common application of ultrasound is size determination. The accuracy of converting a measured organ size to organ mass depends on the location and shape of the organ. For example, a change in mass of the pectoral muscle in birds will result in only a minor change in muscle thickness because this large muscle is rather flat. In general, the accuracy of the size measurement or mass estimation via ultrasound is not as good as by dissection studies, but the variance due to measurement error can be reduced by carrying out multiple measurements (Starck & Burann, 1998; Dietz *et al.*, 1999a; Hassen *et al.*, 1999). This can be achieved by increasing the number of images taken, and in the case of rather vague images, by taking several measurements on the same image. Note that, for each of these replicate treatments, the successive measurements need to be blind and 'independent' in order to really obtain the suggested statistical gains, e.g. Lindström *et al.*, 2000. Thus, the observer should not be able to compare the successive measurements taken in order to completely avoid the chance of subconscious bias (actually, she or he should not be able to see the measurements at all). Independence requires at the minimum that the probe is repositioned at the start of each successive measurement. It is much better to completely separate the repeat measurements in time, e.g. Dietz *et al.*, 1999a. In general terms, we advise testing for measurement error and variance in the sample, and then running a power test to ascertain the number of repeat measurements that are required to answer the particular question in which one is interested.

Applications

Ultrasound scanning can be applied in an enormous range of studies, and the variety of applications will only increase as the ultrasonographic instruments become more mainstream (and therefore much cheaper) and more advanced. It is impossible to comprehensively review all the possible applications in the veterinary sciences, e.g. Hildebrandt *et al.*, (1995, 1998) agricultural and zootechnical sciences, e.g. Perkins *et al.* (1992), Herring *et al.* (1994), ecophysiology, e.g. Haefner *et al.* (1996), Starck & Burann (1998), Dietz *et al.* (1999b) and animal ecology (Stephenson *et al.*, 1998; Piersma *et al.*, 1999). Instead, to convey the great scope of ultrasound applications in zoology and evolutionary biology, we will review a few ultrasound applications with which we are familiar and that we regard as particularly promising and exciting.

Implantation date of embryos in relation to body condition

Recent technological advancements and continued miniaturization of instruments has brought up a second generation of high performance portable ultrasonographs. With such miniaturized equipment for ultrasonography, it has become possible to include exotic and rare animals into field studies monitoring reproductive biology of wild animals, e.g. Woodroffe (1995). The small size of ultrasonographs and independence from energy outlets make such systems suitable for field studies, even under challenging field conditions. Outstanding examples include studies by Hildebrandt *et al.* (1996, 1998) of the reproductive cycle of African elephants in the wild. Based on a capture–recapture programme, a portable ultrasonography system, equipped with a 7.5 MHz linear array scanner head and conducting transrectal ultrasonography, was used to monitor the ovarial cycle in free ranging elephants over a period of 3 years. The programme helped to develop an effective contraception programme in areas of dense overpopulation of African elephants and thus avoided the killing of hundreds of young elephants as usually practised. Continued development in the understanding of ultrasound anatomy of exotic animals and advancements in the skills for handling such large animals have now resulted in worldwide reproductive biology programmes for wild and captive mammals.

Diet switching and gizzard sizes in birds

Birds in general, and shorebirds in particular, show large interspecific variation in the relative size of the muscular part of the stomach, the gizzard (Piersma *et al.*, 1993, 1999; Starck 1999a, b). Much of this variation can be explained by their diet. Shorebird species feeding mainly on hard-shelled prey such as bivalves and gastropods have large stomachs, whereas species feeding on soft-bodied prey such as worms have small stomachs (Piersma *et al.*, 1993). In addition, large-stomached shorebird species show dramatic changes in gizzard size that appear to correlate with migratory events and the associated changes in diet (Piersma *et al.*, 1999). For example, red knots make very long-range flights between their arctic and subarctic alpine breeding grounds and coastal wetlands at temperate to subantarctic latitudes. On the tundra breeding grounds, knots feed on soft-bodied insects and have relatively small stomachs. At the coastal non-breeding sites where they spend 9 of the 12 months of the year, red knots only eat shellfish that are ingested whole. During this time they carry the large muscular gizzards necessary to crush the bivalves (Piersma *et al.*, 1993). By following the movements during foraging of radiotagged red knots of which gizzard size had been determined upon capture using ultrasonography (Dietz *et al.*, 1999a), it was possible to show that individuals that fed on bivalve species that are notoriously difficult to crush (such as the cockle *Cerastoderma edule*), carried heavier stomachs than individuals that fed on lighter prey (Piersma *et al.*, 1999). Hence, interaction between diet and stomach seems to be a dynamic one. Subsequent studies have indeed shown that a soft diet can induce stomach reductions of 50% within a week, and that such changes are reversible (A. Dekinga *et al.* unpublished observations; see also Starck, 1999a, b). In flying animals like these shorebirds, weight-saving must be a strong selective force on form and function. We would thus expect knots to be built in ways where large organs such as the stomach are always kept at minimal size. Knowing the energetic pay-offs of variations in size, i.e. what are the costs involved in building and maintaining large gizzards; what benefits are gained by having a large gizzard in terms of potential intake rate and digestive efficiency, using ultrasound scanning as a central technique it may be possible to carry out true cost–benefit analyses of gizzard size in different ecological contexts.

Studies of phenotypic plasticity and heritability

Ultrasonography as a tool of non-invasive imaging allows us to include new and promising perspectives into laboratory based studies on internal organ size and function. The virtually unlimited number of repeat measurements that one can make on the same individual enables testing of the effects of different environmental conditions on that individual. Following such an approach, it is possible to study not only changes of the phenotype of internal organs in response to a change of external conditions but also dynamics, reversibility and repeatability of such changes of the phenotype. For example, Starck and Burann (1998) and Starck and Beese (2000) have studied size responses of the gastrointestinal tract of snakes over a period of 880 days and 12 feeding intervals. Such studies provide new insights into amount and dynamics of organ size changes under fluctuating external conditions. Duplex ultrasonographic studies gave a clear insight into changing patterns of blood (volume) flow to the intestines and liver depending on the feeding condition of the snake (Fig. 7.2(*c*)). Similar studies have been performed on birds, presenting evidence for an astonishingly flexible phenotype of internal organs (Starck, 1999a, b). In principle, it will be possible to use ultrasonography not only in microevolutionary studies of phenotypic plasticity but also in quantitative genetic breeding designs. Ultrasonography-based morphometry of internal organs, e.g. stomach, intestines, or liver in parents and their offspring, opens conventional quantitative genetic research avenues to the study of internal organs. Such studies are not possible without non-invasive imaging techniques simply because an animal cannot be killed twice. The combination of classical research approaches with new technologies does not only extend the field of investigation, it also gives new perspective into the flexibility of internal organs and the heritable components of such flexibility.

Minimal invasive techniques in combination with ultrasonography

With non-invasive imaging we can study the size of internal organs, blood volume flow, and patterns of change. However, if one is interested in understanding the cytological or biochemical/physiological basis of such changes one needs to obtain tissue samples for detailed analyses. Ultrasonography in combination with microneedle biopsy techniques allows for very precise tissue sampling. In such procedures, ultrasonographs equipped with specific programming facilities, show the biopsy channel on the screen and let the practitioner find the most suitable position for taking the biopsy and to avoid tissue damage and piercing of vessels.

Acknowledgements

JMS is supported by DFG grants STA345/4-1, 4-2, 5-1, 5-2. The work of MWD and TP is supported by a PIONIER grant from the Netherlands Organization for Scientific Research (NWO). We thank Anne Dekinga and Åke Lindström for their stimulating involvement in developing ultrasound techniques in shorebird studies. We thank Barbara Helm and Kathleen Beese for critically reading the manuscript. This is NIOZ-publication 3488.

REFERENCES

Dietz, M.W., Dekinga, A., Piersma, T. & Verhulst, S. (1999a). Estimating organ size in small migrating shorebirds with ultrasonography: an intercalibration exercise. *Physiology, Biochemistry and Zoology*, **72**, 28–37.

Dietz, M.W., Piersma, T. & Dekinga, A. (1999b). Body-building without power training: endogenously regulated pectoral muscle hypertrophy in confined shorebirds. *Journal of Experimental Biology*, **202**, 2831–7.

Göritz F. (1993). Sonographie bei Zoo- und Wildtieren. PhD dissertation, Veterinary Medicine, Free University of Berlin.

Haefner, P.A., Jr. (1996). Applications of ultrasound technology to crustacean physiology; monitoring cardiac and scaphognathite rates in Brachyura. *Crustaceana*, **69**, 788–94.

Haefner, P.A., Jr., Sheppard, B., Barto, J., McNeill, E. & Cappelino, V. (1996). Application of ultrasound technology to molluscan physiology: non-invasive monitoring of cardiac rate in the blue mussel, *Mytilus edulis* Linnaeus, 1758. *Journal of Shellfish Research*, **15**, 685–7.

Hassen, A., Wilson, D.E., Amin, V.R. & Rouse, G.H. (1999). Repeatability of ultrasound-predicted percentage of intramuscular fat in feedlot cattle. *Journal of Animal Science*, **77**, 1335–40.

Hassler, D. (1995). Ultraschalltechnik. In *Bildgebende Systeme für die medizinische Diagnostik*, 3rd edn. ed. H. Morneburg, pp. 191–220. München: Publicis MCD Verlag.

Herring, W.O., Miller, D.C., Bertrand, J.K. & Benyshel, L.L. (1994). Evaluation of machine, technician, and interpreter effects on ultrasonic measurements of backfat and longissimus muscle area in beef cattle. *Journal of Animal Science*, **72**, 2216–26.

Hildebrandt, T., Pitra, C. & Thielebein, J. (1994). Transintestinale Ultraschalluntersuchung bei Nützgeflügel. *Mh. Vet.-Med.*, **49**, 337–43.

Hildebrandt, T., Pitra, C., Sömmer, P. & Pinkowski M. (1995). Sex identification in birds of prey by ultrasonography. *Journal of Zoology and Wildlife Medicine*, **26**, 367–76.

Hildebrandt, T., Göritz, F., Quandt, S. *et al.* (1996). Ultrasonography as a tool to evaluate the reproductive tract in female Asian and African elephants. *Journal of Ultrasound Medicine*, **15**, S59.

Hildebrandt, T.B., Göritz, F., Stetter, M.D., Hermes, R. & Hofmann, R.R. (1998). Applications of sonography in vertebrates. *Zoology*, **101**, 200–9.

Lindström, Å., Kvist, A., Piersma, T., Dekinga, A. & Dietz, M.W. (2000). Avian pectoral muscle size rapidly tracks body mass changes during flight, fasting and fuelling. *Journal of Experimental Biology*, **203**, 913–19.

Marcelis, S., Daenen, B. & Ferrara, M.A. (1996). *Peripheral Musculoskeletal Ultrasound Atlas*, ed. R.F. Dondelinger. Stuttgart: Georg Thieme Verlag.

Morneburg, H. (ed.) (1995). *Bildgebene Systeme für die medizinische Diagnostik*, 3rd edn. München: Publicis MCD Verlag.

Ødegaard, S., Kimmey, M.B., Marin, R.W., Yee, H.C., Cheung, A.H.S. & Silverstein, F.E. 1992. The effects of applied pressure on the thickness, layers, and echogenicity of gastrointestinal wall ultrasound images. *Gastrointestinal Endoscopy*, **38**, 351–6.

Perkins, T.L., Green, R.D., Hamlin, K.E., Shepard, H.H. & Miller, M.F. (1992). Ultrasonic prediction of carcass merit in beef cattle: evaluation of technician effects on ultrasonic estimates of carcass fat thickness and longissimus muscle area. *Journal of Animal Science*, **70**, 2758–65.

Piersma, T., Koolhaas, A. & Dekinga, A. (1993). Interactions between stomach structure and diet choice in shorebirds. *Auk*, **110**, 552–64.

Piersma, T., Dietz, M.W., Dekinga, A. *et al.* (1999). Reversible size-changes in stomachs of shorebirds: when, to which extent, and why? *Acta Ornithologica*, **34**, 175–81.

Soldner, R. (1995). Ultraschalltechnik, Sicherheitsaspekte. In *Bildgebende Systeme für die medizinische Diagnostik*, 3rd edn. ed. H. Morneburg, pp. 221–5. München: Publicis MCD Verlag.

Starck, J.M. (1999a). Phenotypic flexibility of the avain gizzard: rapid, reversible and repeated changes of organ size in response to changes in dietary fibre content. *Journal of Experimental Biology*, **202**, 3171–9.

Starck, J.M. (1999b). Structural flexibility of the gastro-intestinal tract of vertebrates. Implications for evolutionary morphology. *Zoologischer Anzeiger*, **238**, 87–101.

Starck, J.M. & Beese, K. (2000). Structural flexibility of the intestine of Burmese python in response to feeding. *Journal of Experimental Biology*, **24**, in press.

Starck, J.M. & Burann, A-K. (1998). Non-invasive imaging of the gastrointestinal tract of snakes: A comparison of normal anatomy, radiography, magnetic resonance imaging, and ultrasonography. *Zoology*, **101**, 210–33.

Stephenson, T.R., Hundertmark, K.J., Schwartz, C.C. & Van Ballenberghe, V. (1998). Predicting body fat and body mass in moose with ultrasonography. *Canadian Journal of Zoology*, **76**, 717–22.

Urbak, S.F., Pedersen, J.K. & Thorsen, T.T. (1998). Ultrasound biomicroscopy. II. Intraobserver and interobserver reproducibility of measurements. *Acta Ophthalmologica Scandinavica*, **76**, 546–9.

Woodroffe, R. (1995). Body condition affects implantation date in the European badger, *Meles meles*. *Journal of Zoology, London*, **236**, 183–8.

TIMOTHY R. NAGY

8

The use of dual-energy X-ray absorptiometry for the measurement of body composition

Introduction and historical overview

Dual-energy X-ray absorptiometry (DXA) is a rapid, non-invasive technique that allows for the measurement of total and regional bone mineral (TBBM, g), total bone mineral density (TBMD, g/cm^2), fat mass (FM, g) and bone-free lean tissue mass (LM, g) *in vivo* in animals ranging in size from man to mouse. Since its introduction slightly more than a decade ago (Cullum *et al.*, 1989; Mazess *et al.*, 1989; Blake & Fogelman, 1997), DXA has become the most widely used technology for measuring TBBM and TBMD in humans, and its use for determining FM and LM mass is widespread. The use of DXA in animal research is also increasing (Grier *et al.*, 1996). Recent advances in both hardware and software now allow for the determination of body composition in mouse-sized animals (Hunter & Nagy, 1999; Nagy & Wharton, 1999; Nagy *et al.*, 1999a,b; Nagy & Clair, 2000).

Historically, the use of absorptiometry or densitometry for the measurement of TBBM and TBMD began with the advent of single-photon absorptiometry (SPA) in the 1960s (Cameron & Sorenson, 1963; Cameron *et al.*, 1968). Although this technique proved useful, its applicability was limited to measuring bones that could be immersed in water or other soft tissue-like materials (Grier *et al.*, 1996; Blake & Fogelman, 1997). The development of dual-photon absorptiometry (DPA) overcame the limitations of SPA and allowed for the determination of TBBM and TBMD in areas that were not surrounded by homogeneous soft tissue. In addition, the use of a second energy level allowed for the determination of two tissue types simultaneously (Gotfredsen *et al.*, 1986; Heymsfield *et al.*, 1989; Mazess *et al.*, 1990). In the late 1980s, the radionuclide sources used in DPA were replaced with an X-ray source, creating DXA. The use of an

Table 8.1. *Mass attenuation coefficients (μ_m) and R values for elements common to the body*

Element	Atomic weight	μ_m at 40 keV	μ_m at 70 keV	R
Hydrogen	1.008	0.3458	0.3175	1.0891
Carbon	12	0.2047	0.1678	1.2199
Nitrogen	14	0.2246	0.1722	1.3043
Oxygen	16	0.2533	0.1788	1.4167
Calcium	40.1	1.792	0.5059	3.5422

Source: Adapted from Pietrobelli *et al.* (1996).

X-ray source increased resolution and precision, and eliminated the problem of radionuclide decay (Blake & Fogelman, 1997).

Theoretical aspects of DXA

The theoretical aspects of DXA have been described and reviewed numerous times (Cullum *et al.*, 1989; Mazess *et al.*, 1989; Pietrobelli *et al.*, 1996; Blake & Fogelman, 1997). The goal of this section will be to provide the reader with an overview of the basic assumptions and methods associated with the technology.

The basic principle of absorptiometry relies on the attenuation of photons as they pass through an absorber, e.g. FM, LM, TBBM. The amount of attenuation, measured as the fractional lowering of beam intensity, is dependent upon many things, including the initial energy level of the photon, the linear attenuation coefficient and path length. Since the linear attenuation coefficient is density dependent, one can calculate the mass attenuation coefficient by dividing the linear attenuation coefficient by the density of the substance being measured. For any element, the mass attenuation coefficient is constant at a given photon intensity, but decreases with increasing photon intensity (Hubbell, 1982). The use of two energy levels (as with DXA), allows for the calculation of the '*R* value', which is the mass attenuation coefficient at the lower energy level, divided by the coefficient at the higher energy level (Table 8.1). Therefore, it is possible to use DXA to identify an element based upon its unique mass attenuation coefficient and the *R* value.

The concept of *R* values can be used to identify soft tissue (FM and LM) if one assumes that these have constant attenuation values. For instance, the theoretical *R* value for a given fatty acid or triglyceride can be calculated based on the mass fraction of its elemental composition. Pietrobelli *et al.* (1996) calculated that the *R* value for fats averaged 1.21 (using 40 and 70 keV) with very little variance (1.206–1.229). The authors then tested this empirically by mixing lard with ground beef in varying amounts. Extrapolating the regression of *R* against percentage fat to 100% fat gave an empirical *R* value of 1.19. This value is most likely not significantly different from 1.21 due to experimental error and the inability to measure the exact energy level (keV) of the two DXA energies. Similar experiments have determined the *R* values for LM (1.372; Pietrobelli *et al.*, 1998).

Because the major components of the body (FM, LM, and TBBM) are composed of different mass fractions of elements, their calculated and empirically determined *R* values differ. Fat mass, composed mostly of low *R* value elements (carbon, hydrogen, and oxygen), produces the lowest *R* value (1.21). Lean tissue mass contains the elements found in FM, plus salts (sodium, chlorine, magnesium, and potassium). These latter elements are of higher atomic number, higher mass attenuation coefficients, and higher *R* values, and give LM an average *R* value of 1.372. Lastly, TBBM contains carbon, oxygen, salts, and a large amount of calcium. Because calcium has a large *R* value (3.5422) and is abundant in bone, the *R* value for bone mineral is high (2.8617). These differences in the ratio of mass attenuation coefficients between the components of body composition (FM, LM, and TBBM) are the reason that DXA can differentiate tissues.

Now that the basic concept of determining the tissue type has been demonstrated, it must be pointed out that, when scanning an animal with DXA, it is highly unlikely that, at any given measurement point, only one tissue would be present. Instead, the body is a mixture of the three components (FM, LM, and TBBM). How then does DXA determine body composition in this heterogeneous environment?

Prior to addressing this issue, it is important to explain how DXA scans and analyses an animal. For ease of presentation, the scenario of a DXA analysis based on pencil beam technology will be presented. The animal to be scanned is placed in a prostrate position on the table-top of the instrument. An X-ray source, usually mounted in the base of the instrument, provides the photons, and a photomultiplier tube located above the animal is used to quantify the photons. Both the X-ray source and photomultiplier tube move in a rectilinear pattern until the animal has been completely scanned. As the DXA progresses, information is collected as pixels that vary in size depending upon instrument and scan mode (see below). Thus, the information collected on the animal is based upon thousands of pixels of data. This can be visualized by imagining a transparent mouse lying on a piece of graph paper. Each pixel of data then corresponds to a square on the graph paper. Software provided with the instrument then determines the proportion of the different body components associated with each pixel, and sums these to represent the entire animal.

Because DXA uses only two energy levels, only two components can be solved at any one time. Thus, DXA discriminates a pixel as either 'FM and LM' or 'TBBM and soft-tissue'. The differentiation of the pixel is referred to as point-typing and is accomplished using dynamic models that incorporate the pixel *R*-value and the rate of change in *R*-value based on surrounding pixels. Point-typing poses a problem when a pixel is point typed as bone, as no information is available on the proportion of FM and LM that comprises the soft tissue mass in the pixel. To determine the proportion FM and LM in 'bone' pixels, manufacturers have developed models to predict the composition based on the composition of the soft tissue surrounding the bone. These models are usually proprietary, and their assumptions may lead to inaccuracies in the determination of body composition.

The attenuation of the X-ray beams (I) is dependent upon the initial X-ray intensity (I_o), the mass attenuation coefficients of bone (μ_B) and soft-tissue (μ_{ST}) and their area density (M_B and M_{ST}; g/cm²) as shown by Heymsfield *et al.* (1989):

$$I^{LE} = I_o^{LE} \exp^{-[\mu_{ST}^{LE} \times M_{ST} + \mu_B^{LE} \times M_B]} \quad (1)$$

$$I^{HE} = I_o^{HE} \exp^{-[\mu_{ST}^{HE} \times M_{ST} + \mu_B^{HE} \times M_B]} \quad (2)$$

Where LE and HE represent the low and high energy X-rays respectively. Because the mass attenuation coefficients are known from calibration and the X-ray intensity is measured by DXA, the equations can be rearranged to solve for mass (Heymsfield *et al.*, 1989):

$$M_B = \frac{(\mu_{ST}^{LE}/\mu_{ST}^{HE}) \times \ln(I^{HE}/I_o^{HE}) - \ln(I^{LE}/I_o^{LE})}{\mu_B^{LE} - \mu_B^{HE} \times (\mu_{ST}^{LE}/\mu_{ST}^{HE})} \quad (3)$$

$$M_{ST} = \frac{\ln(I^{LE}/I_o^{LE}) - (\mu_B^{LE}/\mu_B^{HE}) \times \ln(I^{HE}/I_o^{HE})}{(\mu_B^{LE}/\mu_B^{HE}) \times \mu_{ST}^{HE} - \mu_{ST}^{LE}} \quad (4)$$

If the pixel contains only soft tissue (FM and LM), DXA is able to determine the proportion of FM and LM using the *R* value (μ^{LE}/μ^{HE}). In this case, the R value of the soft tissue (R_{ST}) will be a function of the proportion of FM (R_{FM}) and LM (R_{LM}) in each pixel, as shown by Gotfredsen *et al.* (1986).

$$\text{Lean tissue fraction} = \frac{(R_{ST} - R_{FM})}{(R_{LM} - R_{FM})} \qquad (5)$$

Solving the equation using known values of R_{FM} and R_{LM} allows for the calculation of FM and LM in that pixel.

In conclusion, DXA utilizes the phenomenon- that elements attenuate photons to different degrees based upon their atomic weights. Since the three major components of the body (FM, LM, and TBBM) are composed of different elements with different attenuation properties, the proportion of each component can be determined for each pixel. However, only two components can be determined within a pixel (FM + LM or TBBM + soft tissue). Thus, in pixels containing bone, the composition of the soft tissue is assumed to be similar to the composition of the soft-tissue surrounding the bone (non-bone pixels). For further information on the theoretical assumptions and workings of DXA, readers are encouraged to see reviews by Heymsfield *et al.* (1989), Gotfredsen *et al.* (1986), Pietrobelli *et al.* (1996), Grier *et al.* (1996), and Blake and Fogelman (1997).

Practical aspects of DXA usage

From a practical standpoint, the major utility of DXA is the ability to measure *in vivo* body composition, including TBBM and TBMD. Although there are many *in vivo* methods for measuring fat and lean tissue in animals (see other chapters in this book), no other single method can provide information on TBBM and TBMD.

The major use for DXA has been for research in the area of osteoporosis, and more recently, obesity. Therefore, the majority of DXA instruments are large enough to accommodate humans up to approximately 150 kg, and have a scan area of approximately 195 by 66 cm (Fig. 8.1). The use of DXA for determining FM and LM in large animals has not received as much study. These studies have used clinical machines and were aimed primarily at using swine to validate DXA use in human studies. However, bone mineral content and TBMD have been determined in a number of large animals, including swine (Mosekilde *et al.*, 1993), sheep (Turner *et al.*, 1995a, b, c; Grier *et al.*, 1996), primates (Jayo *et al.*, 1991, 1994; Colman *et al.*, 1999), and dogs (Drezner & Nesbitt, 1990; Puustjärvi *et al.*, 1992; Grier *et al.*, 1996).

Fig. 8.1. Clinical 'table-top' dual-energy X-ray absorptiometer (Lunar DPX-L; photo courtesy of Lunar Inc.). Although mostly used for the analysis of human body composition, this sized instrument has been used to measure everything from pigs to dogs, cats, and large rats.

In addition to the use of clinical machines for analysis of large animals, software modifications have allowed their application to small animal measurements. The three major manufacturers of DXA instruments (Hologic, Inc., Bedford, MA; Lunar Corp., Madison, WI; and Norland Medical Systems, Inc., Fort Atkinson, WI) all produce 'pediatric' software that modifies pixel size to allow the instrument to be used with the pediatric population. These modifications have permitted smaller animals, such as minipigs (Koo *et al.*, 1995; Pintuaro *et al.*, 1996) and small primates (*Macaca mulatta*; Lane *et al.*, 1995) to be analysed using DXA. In addition, DXA manufacturers have recently produced 'small-animal' software that allows determination of body composition in animals ranging from 0.15 to 5 kg.

In addition to clinical 'table-top' densitometers, peripheral DXA instruments have been developed to measure TBBM and TBMD in the forearm and heel of humans. Both of these instruments are now available in small-animal versions, and allow for the determination of body composition in 'mouse-sized' animals (Lunar PIXImus and Norland pDEXA Sabre; Fig. 8.2). These instruments are relatively small (30×33×63 and 52×43×42.5 cm, respectively), weigh approximately 27 kg, are portable, and run on conventional electrical current. To date, they have not been used in field research, but clearly this potential exists, as long as an electri-

Fig. 8.2. Peripheral dual-energy X-ray absorptiometers for the measurement of 'mouse-sized' animals: the Lunar PIXImus (left) and the the Norland pDEXA Sabre (right). (Photo by T. Nagy.) These instruments are relatively small ($30 \times 33 \times 63$ and $52 \times 43 \times 42.5$ cm, respectively), weigh approximately 27 kg, are portable, and run on conventional electrical current. To date, they have not been used in field research, but clearly this potential exists, as long as an electrical source is available.

cal source is available. Typical scans from these machines are presented in the Appendix.

One limiting factor of the peripheral densitometers described above is their relatively small imaging area. The Lunar PIXImus densitometer has an image area of 80×65 mm, which severely restricts the size of animal that can be scanned accurately. Even moderately sized mice (30 g) cannot be scanned completely. To accommodate this limitation, we have adopted the technique of excluding the head from the analysis. By placing the head outside of the imaging area, we are able to scan obese mice weighing up to 65 g. During analysis of the acquired data, the operator places an exclusion

Table 8.2. *Available sample size and approximate scan speeds for the Lunar DPX-L using the small animal software package*

Mode/speed	Animal weight (kg)	Sample (pixel) size (mm)	Scan time[a] (min)
Detail/medium	4–5	1.2×2.4	5
Detail/slow	0.5–3	1.2×2.4	9
HiRes/medium	0.4–0.5	0.6×1.2	18
HiRes/slow	<0.4	0.6×1.2	34

Notes:
[a] Scan time is based on scanning an area of 120×190 mm. Actual scan time will vary depending upon size of the animal.
Source: Data taken from the *Lunar Small Animal Software Operator's Manual.*

region of interest over the head, and body composition is determined for the postcranial body. Animals' heads were subsequently removed before conducting chemical carcass analysis for the purpose of validating the densitometric determination of body composition (see below). The imaging area of the pDEXA Sabre (133×108 mm) is slightly larger than that of the PIXIMUS. The increased area has allowed us to measure body composition of collared lemmings (*Dicrostonyx groenlandicus*) weighing as much as 72 g (Hunter & Nagy, 1999).

The ability to measure *in vivo* body composition is clearly a valuable tool for researchers. However, one must keep in mind that the ability to conduct these measurements is dependent upon keeping the animal motionless during the scanning process. Thus, animals need to be anaesthetized during the scan and total scanning time becomes an important factor when determining feasibility. Clearly, fast scan times are preferable, but there is usually a trade-off between speed and reliability on a given DXA instrument. For pencil-beam densitometers, the time needed to conduct a scan will depend upon two factors: the sample size (resolution or pixel size) and the sample interval (time at each sampling point or speed). Many DXA manufacturers allow the user to choose both pixel resolution and scan speeds. An example of the selectable modes for the Lunar DPX-L are given in Table 8.2. The Norland pDEXA Sabre also allows the operator to select a variety of resolutions (0.1×0.1 to 1.0×1.0 mm) and scan speeds (1 to 40 mm/s). In order to achieve the best results, it is essential to pick the finest resolution possible (see below) and the

Table 8.3. *Approximate scan times for a variety of animals*

Animals and weights	Scan time (min)	Instrument
Swine (15–25 kg)	18–25	Lunar DPX-L, pediatric software
Rats (350–500 g)	25–35	Lunar DPX-L, HiRes/slow, small-animal software
Lemmings (25–72 g)	20–30	Norland pDEXA Sabre (0.5 × 0.5 mm at 8 mm/s)
Mice (20–35 g)	18–25	Norland pDEXA Sabre (0.5 × 0.5 mm at 8 mm/s)
Mice (10–55 g)	<5	Lunar PIXIMUS (0.18 × 0.18 mm)

slowest scan speed, keeping in mind the total amount of time that your experimental animal can be safely anaesthetized (see Table 8.3 for some examples of scan times). Software limitations with both instruments also constrain the total number of lines (rows or pixels) that can be scanned (Lunar = 410 lines, Norland = 256 lines). Thus, in order to scan a collared lemming that is 125 mm in total length, the resolution setting on the Norland pDEXA Sabre cannot be greater than 0.5 × 0.5 mm (0.5 mm × 256 = 128 mm total scan length).

The Lunar PIXIMUS achieves a relatively rapid scan time by employing cone beam technology (Table 8.3). In essence, the entire image area is illuminated by the X-ray source, and photon attenuation is measured using a stationary detector with a pixel resolution of 0.18 × 0.18 mm. Although this technology allows for rapid and precise measurement of body composition, its limited scan area is a liability.

Availability of DXA

As mentioned previously, there are three major manufacturers of DXA instruments. Each produce multiple 'table-top' DXAs capable of measuring body composition in animals from rats to man. The table-top instruments are expensive, and range from US$35 000 to greater than US$100 000. Peripheral DXA instruments for measuring body composition in mice sell for approximately US$40 000. However, because of the widespread use of DXA for clinical applications, it may be possible for researchers to use existing instruments found in osteoporosis clinics, hospitals, and physician offices. Although it is doubtful that existing equipment will have the small animal software installed or available, the software can be purchased from the manufacturers for less than US$5000.

Validation of DXA

Although the precision and accuracy of DXA as a method for measuring body composition in humans has been critically analysed more than any other technique (Pietrobelli *et al.*, 1996), its use in humans has never been validated by comparison with chemical extraction techniques. Instead, three indirect methods have been employed. The first has been to compare DXA results to those of other techniques used in the measurement of human body composition such as hydrodensitometry, hydrometry, bioelectrical impedance, anthropometry (see Lohman, 1996 for a review). The second has been to place substances on humans (strips of fat or lard, water, ground beef), with analysis before and after, to examine how DXA classified the additional tissue (Snead *et al.*, 1993; Svedsen *et al.*, 1993; Milliken *et al.*, 1996). The third has been to compare DXA-derived data from swine with chemical extraction data, with the assumption that pigs are a good model for humans (Brunton *et al.*, 1993, 1997; Ellis *et al.*, 1994; Pintuaro *et al.*, 1996). Dual-energy X-ray absorptiometry has also been used to measure body composition of swine for agricultural studies (Brunton *et al.*, 1993, 1997; Ellis *et al.*, 1994; Mitchell *et al.*, 1996, 1998a, b; Pintuaro *et al.*, 1996; Elowsson *et al.*, 1998). These studies have, for the most part, found very similar results; body composition values from DXA differ significantly from those obtained from chemical extraction techniques. However, the DXA-derived and chemically derived data are closely related. Thus, regression equations can be developed to predict body composition by chemical extraction from DXA-derived values.

Ellis *et al.* (1994) examined the precision and accuracy of the Hologic QDR-2000 densitometer (Adult Whole-Body software versions 5.56 and 5.57) for measuring body composition in pigs weighing from 5 to 35 kg. Precision was determined using four pigs that were scanned daily for 10 days. The coefficient of variation (CV%) ranged from a low of 1.3% for TBMD to a high of 3.2% for FM. Comparison of DXA to total carcass chemical analysis revealed that DXA underestimated TBBM by 25%. Depending upon the software version, DXA either underestimated or overestimated FM by 19.5 or 15.5% and LM by 8 or 6%. Although these differences were significant, DXA-derived values could be used to accurately predict body composition as determined by chemical analysis. The variance explained by the regression equations ranged from a low of 98.7% for fat and TBBM to a high of 99.8% for LM.

Pintauro *et al.* (1996) cross-calibrated FM and LM as measured by DXA (Lunar DPX-L, adult fast-detail mode, version 1.3 y, and the pediatric medium mode, version 1.5 d) to pig chemical carcass analysis in the pediatric weight range (16–36 kg). Data obtained using the pediatric software showed that DXA overestimated FM, and underestimated LM and TBBM, relative to chemical analysis. Regression equations were then derived to predict chemical LM and FM using DXA. Both models included the body composition parameter of interest (either LM or FM) and body weight. The variance explained by the models was 99.6% and 99.8% for LM and FM, respectively. In addition, duplicate scans were used to determine the reliability of the instrument. The CV's for TBBM, FM, and LM were 1.8, 4.1, and 1.0%, respectively.

The use of DXA for measuring FM, LM, and TBBM of partial carcasses for the meat industry has also been explored. Mitchell *et al.* (1998b) measured 181 pig half-carcasses ranging in weight from 10–51 kg. The head and viscera were removed at slaughter and the remaining carcass was split at the midline. Carcass halves were analyzed using the Lunar DPX-L and the software version used in the analysis was determined by sample weight (pediatric mode if <30 kg; adult mode if $>$ than 30 kg). Data obtained from the DXA analysis were corrected using regression equations from a previous study (Mitchell *et al.*, 1998b). The corrected data were not significantly different from the data obtained via chemical carcass analysis.

Dual-energy X-ray absorptiometry has also been used to measure body composition in dogs and cats. Speakman *et al.* (2001) analysed ten cats and six dogs ranging in weight from 1.8 to 22.1 kg using a Hologic QDR-1000 W densitometer. DXA results were closely correlated with those from chemical analysis (LM $r=0.999$, FM $r=0.992$), and across all animals the absolute and percentage discrepancies were small (LM 119.4 g, 2.64%; FM 28.5 g, 2.04%). However, individual errors were much greater and ranged from underestimates of 2.46 and 20.7% to overestimates of 13.33 and 31.5% for LM and FM, respectively.

The above validation studies were conducted on clinical, 'table-top' densitometers. These instruments are also capable of analysing body composition in smaller animals using 'small-animal' software. Laboratory rats have received the most attention with regard to DXA use in small animals. Numerous studies have validated DXA for measurement of TBBM in rats with good success (Griffin *et al.*, 1993; Lu *et al.*, 1994; Mitlak *et al.*, 1994; Sievänen *et al.*, 1994; Rozenberg *et al.*, 1995; Grier *et al.*,

1996; Keenan *et al.*, 1997). Fewer studies have validated DXA for assessing FM and LM in rats (Jebb *et al.*, 1996; Makan *et al.*, 1997; Bertin *et al.*, 1998; Rose *et al.*, 1998). To my knowledge, published validations exist using Hologic instruments only, even though all of the major DXA manufacturers produce small animal software. Two validation studies have been published using the Hologic QDR-1000W. Jebb *et al.* (1996) measured 12 rats in triplicate without repositioning between scans (software version 5.61P). The CV for the repeated measures for FM and TBBM was less than 1%, no information was presented on LM repeatability. Comparison of DXA data to that of chemical carcass analysis revealed that DXA overestimated FM. On average, TBBM did not differ between methods, although DXA tended to overestimate TBBM at low values and underestimate TBBM at higher values. Data from DXA was well correlated with chemical analysis ($r = 0.96$ for both FM and TBBM). Rose *et al.* (1998) using a later version of the software (version 5.71P) also found that DXA overestimated percentage body fat. Again, DXA was highly related to chemical carcass analysis with correlations of 0.99, 0.96, and 0.81 for FM, LM, and TBBM, respectively.

Bertin *et al.* (1998) performed precision and accuracy tests of the Hologic QDR 4500 (software version V8–19a) for use with rats. Twenty-one rats weighing from 130–468 g were scanned a minimum of three times (with repositioning between scans), resulting in CVs of 1.7, 1.1, and 4.7 for TBBM, LM, and FM, respectively. Accuracy was tested in an additional 26 animals, also scanned in triplicate. As seen previously, DXA overestimated FM, but FM estimates were closely related ($r^2 = 0.97$) to FM determined by chemical analysis. Because the relationship between DXA and chemical extraction was linear with a zero intercept, it was possible to obtain a correction factor of 0.75. After correcting DXA-derived values of percentage FM using the correction factor, the difference between the means (corrected DXA and chemical extraction data) was $0.04 \pm 1.6\%$. An important finding of the study was that there was no significant difference between a single scan and the average of the three scans for animals above 200 g. However, this was not the case for animals weighing less than 200 g, suggesting that 200 g may be close to the lower limit of the instrument and software version. These data suggest that DXA can be used to measure body composition in rats as long as the DXA data is corrected based on chemical extraction data.

Only one study has reported using a 'table-top' densitometer to measure mice. Klein *et al.* (1998) used the Hologic QDR 1500 (mouse

Table 8.4. *Regression equations to predict chemical extraction values using DXA-derived data for TBBM, fat mass (FM), and bone-free lean tissue mass (LM) in collared lemmings* (Dicrostonyx groenlandicus)

Dependent variable	Regression equations[a]	Model r^2	SEE (g)	Probability
Ash (g)	$y = 0.04 + 0.82(\text{TBBM}) - 0.03(\text{FM}) + 0.02(\text{LM})$	0.83	0.12	<0.001
FM (g)	$y = -3.00 - 7.44(\text{TBBM}) - 0.76(\text{FM}) + 0.40(\text{LM})$	0.97	1.06	<0.001
LM (g)	$y = 4.37 - 8.62(\text{TBBM}) + 0.24(\text{FM}) + 0.59(\text{LM})$	0.98	1.28	<0.001

Notes:
[a] TBBM, fat, and lean values derived from DXA (g).

whole body software version 3.2) to measure TBBM in mice. The precision of mouse TBBM measurements was stated to be 0.99% (CV), but no information was given concerning the number of mice used or the number of times each mouse was scanned. Total body bone mineral was highly related ($r^2 = 0.96$) to chemically determined whole body bone calcium in a sample of 12 mice weighing from 10–24 g. Whether the instrument and software are capable of measuring FM and LM in mice was not stated.

Recently, the ability to measure body composition in 'mouse-sized' animals has become possible using peripheral DXA. We have measured body composition of collared lemmings (*Dicrostonyx groenlandicus*) using the Norland pDEXA Sabre system (Nagy *et al.*, 1999a). Male lemmings ($n = 32$) weighing from 25 to 72 g, were fasted for 3 hours, killed by overdose, and scanned twice with repositioning between scans (8 mm/s, 0.5 $\times$0.5 mm resolution, software ver. 3.6). Our results showed that the coefficient of correlation and the mean intra-individual coefficient of variation (CV) for the repeated DXA analysis were: FM: $r = 0.91$, CV = 10.0%; LM: $r = 0.93$, CV = 5.8%, and TBBM: $r = 0.86$, CV = 7.2%. DXA-derived values overestimated FM (3.93 ± 0.71 g) and underestimated LM (6.27 ± 0.77 g) and TBBM (0.14 ± 0.04 g) relative to the carcass analysis ($P < 0.01$). When data from the DXA analysis were used to predict the gravimetric results, the r^2 for FM, LM, and TBBM were 0.97, 0.98, and 0.83, respectively (Table 8.4).

The Lunar PIXImus densitometer has been used to determine body composition in mice (Nagy & Clair, 2000). Twenty-five male C57BL/6J mice (6–11wks; 19–29 g) were anaesthetized and scanned three times (with repositioning between scans) using software version 1.42.006.010. The mean intraindividual CV for the repeated DXA analyses were: TBBM, 1.60%; FM, 2.20%, and LM, 0.86%. Accuracy was determined by comparing the DXA-derived data from the first scan with the chemical analysis data (Fig. 8.3). DXA accurately measured bone ash mass ($P = 0.94$), underestimated LM (0.59 ± 0.05g, $P < 0.001$), and overestimated FM (2.19 ± 0.06g, $P < 0.001$). Thus, DXA estimated 100% of bone ash, 97% of carcass LM, and 209% of carcass fat. DXA-derived values were used to generate predictive equations (or calibration curves) for chemical values using multiple regression techniques. Chemically extracted FM was best predicted by DXA FM and DXA LM (model $r^2 = 0.86$), and chemically determined LM by DXA LM ($r^2 = 0.99$). Taken together, these data suggest that DXA can be used to determine body composition in animals.

Conclusion

Dual-energy X-ray absorptiometry is a non-invasive technique capable of measuring body composition (TBBM, TBMD, FM, and LM) in animals ranging from 0.01 to 150 kg. Although the technique has mostly been used for clinical purposes, its use for animal research is increasing. One major drawback is the cost of the instrument, which can range from US$35 000 to greater then US$100 000.

The use of DXA has been validated against chemical carcass analysis for a variety of organisms. In most cases, the body composition results obtained with DXA are significantly different from chemical analysis. However, the relationship between DXA and chemical analysis is strong, allowing DXA results to be corrected using regression techniques.

REFERENCES

Bertin, E., Ruiz, J-C., Mourot, J., Peiniau, P. & Portha, B. (1998). Evaluation of dual-energy X-ray absorptiometry for body-composition assessment in rats. *Journal of Nutrition*, **128**, 1550–4.

Blake, G.M. & Fogelman, I. (1997). Technical principles of dual energy X-ray absorptiometry. *Seminar in Nuclear Medicine*, **27**, 210–28.

Brunton, J.A., Bayley, H.S. & Atkinson, S.A. (1993). Validation and application of dual-energy X-ray absorptiometry to measure bone mass and body composition in small infants. *American Journal of Clinical Nutrition*, **58**, 839–45.

Brunton, J.A., Weiler, H.A. & Atkinson, S.A. (1997). Improvement in the accuracy of dual energy X-ray absorptiometry for whole body and regional analysis of body composition: validation using piglets and methodologic considerations in infants. *Pediatric Research*, **41**, 590–6.

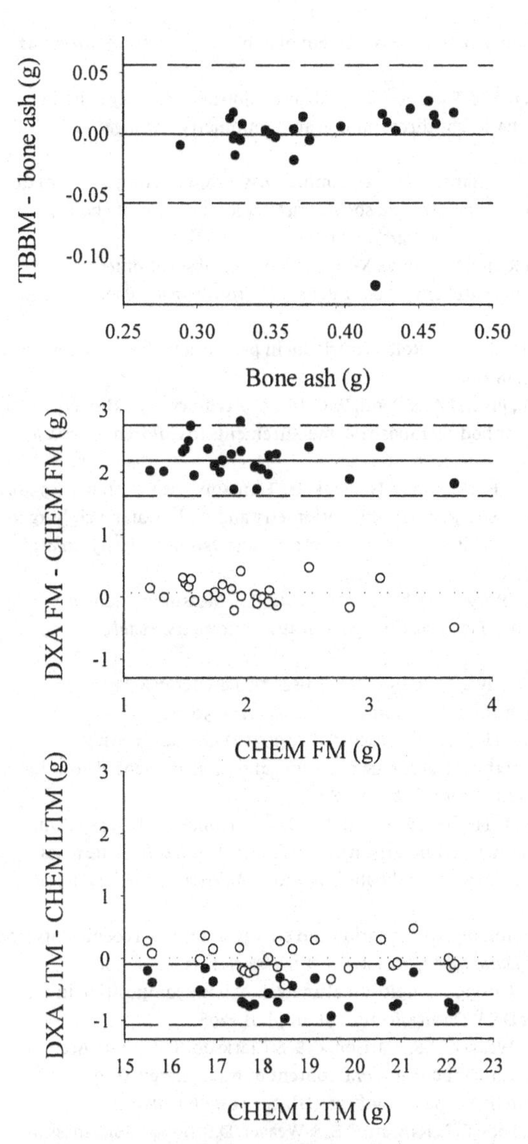

Fig. 8.3. Residual plots of the difference between methods vs. the criterion method (chemical carcass analysis; CHEM). Filled circles represent unadjusted data with the solid line being the mean difference between the methods. Large dashed lines represent the mean ± 2 s.d. Open circles are the adjusted (predicted) values using the following regression equations: CHEM FM (g) = $-0.50 + 1.09$ (DXA FM) $-$ 0.11(DXA LM), $r^2 = 0.86$; CHEM LM (g) = $-0.14 +$ 1.04(DXA LM), $r^2 = 0.99$. The small-dashed line represents the mean difference between predicted and observed. (Reprinted with permission from Nagy & Clair, 2000.)

Cameron, J.R. & Sorenson, J. (1963). Measurement of bone mineral *in vivo*. *Science*, **42**, 230–2.

Cameron, J.R., Mazess, R.B. & Sorenson, J. (1968). Precision and accuracy of bone mineral determination by direct photon absorptiometry. *Investigative Radioliology*, **3**, 141–50.

Colman, R.J., Hudson, J.C., Barden, H.S. & Kemnitz, J.W. (1999). A comparison of dual-energy X-ray absorptiometry and somatometrics for determining body fat in Rhesus macaques. *Obesity Research*, **7**, 90–6.

Cullum, I.D., Ell, P.J. & Ryder, J.P. (1989). X-ray dual photon absorptiometry: a new method for the measurement of bone density. *British Journal of Radiology*, **62**, 587–92.

Drezner, M.K. & Nesbitt, T. (1990). Role of calcitriol in prevention of osteoporosis: Part 1. *Metabolism*, **39**, 18–23.

Ellis, K.J., Shypailo, R.J., Pratt, J.A. & Pond, W.G. (1994). Accuracy of dual-energy x-ray absorptiometry for body composition measurements in children. *American Journal of Clinical Nutrition*, **60**, 660–5.

Elowsson, P., Forslund, A.H., Mallmin, H., Feuk, U., Hansson, I. & Carlsten, J. (1998). An evaluation of dual-energy X-ray absorptiometry and underwater weighing to estimate body composition by means of carcass analysis in piglets. *Journal of Nutrition*, **128**, 1543–9.

Gotfredsen, A., Jensen, J. Borg, J. & Christiansen, C. (1986). Measurement of lean body mass and total body fat using dual photon absorptiometry. *Metabolism*, **35**, 88–93.

Grier, S.J., Turner, A.S. & Alvis, M.R. (1996). The use of dual-energy X-ray absorptiometry in animals. *Investigative Radiology*, **31**, 50–62.

Griffin, M.G., Kimble, R., Hopfer, W. & Pacifici, R. (1993). Dual-energy X-ray absorptiometry of the rat: accuracy, precision, and measurement of bone loss. *Journal of Bone Mineral Research*, **8**, 795–800.

Heymsfield, S.B., Wang, J., Heshka, S., Kehayias, J.J. & Pierson, R.N. Jr. (1989). Dual-photon absorptiometry: comparison of bone mineral and soft tissue mass measurements *in vivo* with established methods. *American Journal of Clinical Nutrition*, **49**, 1283–9.

Hubbell, J.H. (1982). Photon mass attenuation and energy-absorption coefficients from 1 keV to 20 MeV. *Journal of Applied Radiation and Isotopes*, **33**, 1269–90.

Hunter, H.L. & Nagy, T.R. (1999). Longitudinal analysis of body composition in collared lemmings using DXA. *Obesity Research*, 7(Suppl. 1), 129S.

Jayo, M.J., Rankin, S.E., Weaver, D.S., Carlson, C.S. & Clarkson, T.B. (1991). Accuracy and precision of lumbar bone mineral content by dual-energy X-ray absorptiometry in live female monkeys. *Calcified Tissue International*, **49**, 438–40.

Jayo, M.J., Jerome, C.P., Lees, C.J., Rankin, S.E. & Weaver, D.S. (1994). Bone mass in female cynomolgus macaques: a cross-sectional and longitudinal study by age. *Calcified Tissue International*, **54**, 231–6.

Jebb, S.A., Garland, S.W., Jennings, G. & Elia, M. (1996). Dual-energy X-ray absorptiometry for the measurement of gross body composition in rats. *British Journal of Nutrition*, **75**, 803–9.

Keenan, M.J., Hegsted, M., Jones, K.L. *et al.* (1997). Comparison of bone density measurement techniques: DXA and Archimedes' principle. *Journal of Bone Mineral Research*, **12**, 1903–7.

Klein, R.F., Mitchell, S.R., Phillips, T.J., Belknap, J.K. & Orwoll, E.S. (1998). Quantitative trait loci affecting peak bone mineral density in mice. *Journal of Bone Mineral Research*, **13**, 1648–56.

Koo, W.W.K., Massom, L.R. & Walters, J. (1995). Validation of accuracy and precision of dual energy X-ray absorptiometry for infants. *Journal of Bone Mineral Research*, **10**, 1111–15.

Lane, M.A., Reznick, A.Z., Tilmont, E.M. *et al.* (1995). Aging and food restriction alter some indices of bone metabolism in male rhesus monkeys (*Macaca mulatta*). *Journal of Nutrition*, **125**, 1600–10.

Lohman, T.G. (1996). Dual energy X-ray absorptiometry. In *Human Body Composition*, ed. A.F. Roche, S.B. Heymsfield & T.G. Lohman, pp. 63–78. Champaign, IL: Human Kinetics.

Lu, P.W., Briody, J.N., Howman-Giles, R., Trube, A. & Cowell, C.T. (1994). DXA for bone density measurements in small rats weighing 150–250 grams. *Bone*, **15**, 199–202.

Makan, S., Bayley, H.S. & Webber, C.E. (1997). Precision and accuracy of total body bone mass and body composition measurements in the rat using X-ray-based dual photon absorptiometry. *Canadian Journal of Physiology and Pharmacology*, **75**, 1257–61.

Mazess, R., Collick, B., Trempe, J., Barden, H. & Hanson, J. (1989). Performance evaluation of a dual-energy X-ray bone densitometer. *Calcified Tissue International*, **44**, 228–32.

Mazess, R.B., Burden, H.S., Bisek, J.P. & Hanson, J. (1990). Dual-energy x-ray absorptiometry for total-body and regional bone-mineral and soft-tissue composition. *American Journal of Clinical Nutrition*, **51**, 1106–12.

Milliken, L.A., Going, S.B. & Lohman, T.G. (1996). Effects of variations in regional composition on soft tissue measurements by dual-energy X-ray absorptiometry. *International Journal of Obesity*, **20**, 677–82.

Mitchell, A.D., Conway, J.M. & Scholz, A.M. (1996). Incremental changes in total and regional body composition of growing pigs measured by dual-energy X-ray absorptiometry. *Growth, Development and Aging*, **60**, 96–105.

Mitchell, A.D., Scholz, A.M. & Conway, J.M. (1998a). Body composition analysis of small pigs by dual-energy X-ray absorptiometry. *Journal of Animal Science*, **76**, 2392–8.

Mitchell, A.D., Scholz, A.M., Pursel, V.G. & Evock-Glover, C.M. (1998b). Composition analsis of pork carcasses by dual-energy X-ray absorptiometry. *Journal of Animal Science*, **76**, 2104–14.

Mitlak, B.H., Schoenfeld, D. & Neer, R.M. (1994). Accuracy, precision, and utility of spine and whole-skeleton mineral measurements by DXA in rats. *Journal of Bone Mineral Research*, **9**, 119–26.

Mosekilde, L., Weisbrode, S.E., Safron, J.A. *et al.* (1993). Calcium-restricted ovariectomized Sinclair S-1 minipigs: an animal model of osteopenia and trabecular plate perforation. *Bone*, **14**, 379–82.

Nagy, T.R. & Clair, A-L. (2000). Precision and accuracy of dual-energy X-ray absorptiometry for determining *in vivo* body composition of mice. *Obesity Research*, **8**, 392–8.

Nagy, T.R. & Wharton, D. (1999). Precision and accuracy of *in vivo* bone mineral measurements of mouse femurs using DXA. *Journal of Bone Mineral Research*, **14**(Suppl.1), S493.

Nagy, T.R., Onorato, D.P., Jiao, X., Goran, M.I. & Gower, B.A. (1999a). Validation of dual-energy X-ray absorptiometry for the assessment of body composition. *American Zoologist*, **38**, 126A.

Nagy, T.R., Wharton, D., Blaylock, M. & Powell, S. (1999b). Precision and accuracy of *in vivo* bone mineral measurements of mice using dual-energy X-ray absorptiometry. *FASEB Journal*, **13**, A912.

Pietrobelli, A., Formica, C., Wang, Z. & Heymsfield, S.B. (1996). Dual-energy X-ray absorptiometry body composition model: review of physical concepts. *American Journal of Physiology: Endocrinology and Metabolism*, **271**, E941–51.

Pietrobelli, A., Wang, Z., Formica, C. & Heymsfield, S. (1998). Dual-energy X-ray absorptiometry: fat estimation errors due to variation in soft tissue hydration. *American Journal of Physiology: Endocrinology and Metabolism*, **274**, E808–16.

Pintuaro, S.J., Nagy, T.R., Duthie, C.M. & Goran, M.I. (1996). Cross-calibration of fat and lean measurements by dual-energy X-ray absorptiometry to pig carcass analysis in the pediatric body weight range. *American Journal of Clinical Nutrition*, **63**, 293–8.

Puustjärvi, K., Karjalainen, P., Nieminen, J. *et al.* (1992). Endurance training associated with slightly lowered serum estradiol levels decreases mineral density of canine skeleton. *Journal of Bone Mineral Research*, **7**, 619–24.

Rose, B.S., Flatt, W.P., Martin, R.J. & Lewis, R.D. (1998). Whole body composition of rats determined by dual energy X-ray absorptiometry is correlated with chemical analysis. *Journal of Nutrition*, **128**, 246–50.

Rozenberg, S., Vandromme, J., Neve, J. *et al.* (1995). Precision and accuracy of *in vivo* bone mineral measurements in rats using dual-energy X-ray absorptiometry. *Osteoporosis International*, **5**, 47–53.

Sievänen, H., Kannus, P. & Järvinen, M. (1994). Precision of measurement by dual-energy X-ray absorptiometry of bone mineral density and content in rat hindlimb *in vitro*. *Journal of Bone Mineral Research*, **9**, 473–8.

Snead, D.B., Birge, S.J. & Kohrt, W.M. (1993). Age-related differences in body composition by hydrodensitometry and dual-energy X-ray absorptiometry. *Journal of Applied Physiology*, **74**, 770–5.

Speakman, J.R., Booles, D. & Butterwick, R. (2001). Validation of dual energy X-ray absorptiometry (DXA) by comparison to chemical analysis in dogs and cats. *International Journal of Obesity*, **25**, 439–47.

Svedsen, O.L., Haarbo, J., Hassager, C. & Christiansen, C. (1993). Accuracy of measurements of total-body soft-tissue composition by dual energy X-ray absorptiometry *in vivo*. In *Human Body Composition*, ed. K.J. Ellis & J.K. Eastman, pp. 381–3. New York: Plenum.

Turner, A.S., Alvis, M.R., Mallinckrodt, C.H. & Bryant, H.U. (1995a). Dose–response effects of estradiol on bone mineral density in ovariectomized ewes. *Bone*, **17**(Suppl.4), 421S–7S.

Turner, A.S., Alvis, M.R., Mallinckrodt, C.H. & Bryant, H.U. (1995b). Dual-energy x-ray absorptiometry in sheep: experiences with *in vivo* and *in vitro* studies. *Bone*, **17**(Suppl.4), 381S–7S.

Turner, A.S., Alvis, M., Myers, W. & Lundy, M.W. (1995c). Changes in bone mineral density and bone-specific alkaline phosphatase in ovariectomized ewes. *Bone*, **17**(Suppl. 4), 395S–402S.

Appendix 8.1

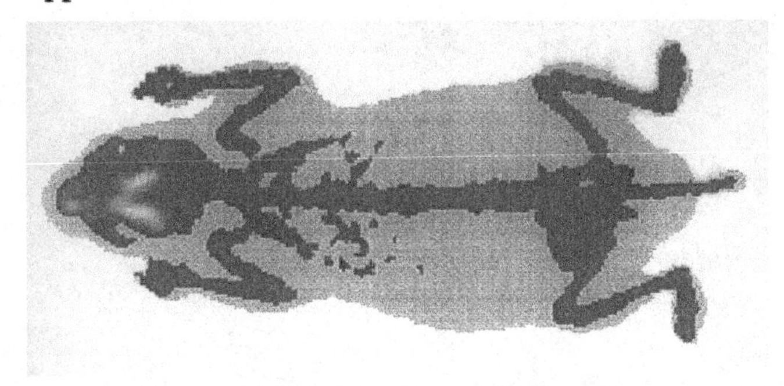

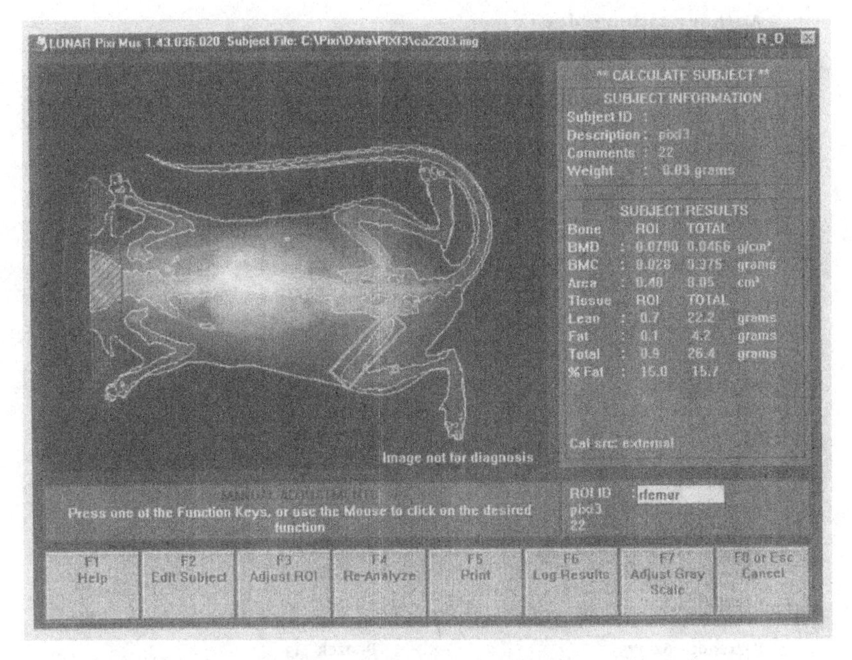

Screen-captured images of a mouse using the Lunar PIXIMUS (*a*) and of a collaraed lemming using the Norland pDEXA Sabre (*b*). In *a*, the head region of the mouse is excluded from the analysis using the oval exclusion region of interest and body composition data as shown under the 'total' column. The 'RQI' column shows the bone mineral data for the rectangular inclusion region of interest that was placed around the animal's right femur.

Index

Author citation index

Includes all authors cited in the text (excluding tables). First author only is cited for papers where there were three or more authors, but both cited where only two. Multiple persons bearing the same surname are not distinguished. Where a range of pages is quoted, the author is cited on all the intervening pages (inclusive). References to authors of chapters in this volume do not include the title pages of their contributions to this work nor the chapter index.

Taxon index

Includes all taxa cited in the text (excluding tables). Latin names are only included here where they were also included in the text. Where page ranges are listed, the taxon is referred to on all the intervening pages (inclusive).

Subject index

Includes all subjects referred to in the main body of text (excluding figures and tables). Where ranges are listed, the subject is referred to on all the intervening pages (inclusive). Some subjects are subdivided and the secondary division is shown in parentheses.